# Number theory

## Ramanujan Mathematical Society Lectures Notes Series


Vol. 1: Number theory

Vol. 2: The Riemann zeta function and related themes

Vol. 3: Formal language aspects of natural computing

Vol. 4: Commutative algebra and combinatorics

Vol. 5: Convexity in discrete structures

Vol. 6: Number theory and discrete geometry

Vol. 7: Discrete mathematics

Vol. 8: Lectures on operator theory

Vol. 9: Essays on geometric group theory

Vol. 10: Techmüller theory and moduli problems

*Ramanujan Mathematical Society*

Lecture Notes Series

Volume 1

# Number theory

Proceedings of the international conference held at the Institute of Mathematical Sciences, Chennai, January 2002

*Volume editors*

S. D. Adhikari
R. Balasubramanian
K. Srinivas

International Press
www.intlpress.com

Ramanujan Mathematical Society
Lecture Notes Series, Volume 1
Number theory

Volume Editors:
S. D. Adhikari
R. Balasubramanian
K. Srinivas



Published in 2010 by International Press, Somerville, Massachusetts, U.S.A. under license from the Ramanujan Mathematical Society.

This work is also published in the Republic of India, exclusively by the Ramanujan Mathematical Society.



ISBN 978-1-57146-186-5

Typeset using the LaTeX system.
Printed in the United States of America.

# Preface

This volume represents the proceedings of the international conference *Analytic Number Theory with special emphasis on L-functions* held at The Institute of Mathematical Sciences, Chennai, in January 2002. Some invited speakers who could not accept our invitation have kindly contributed to this volume. On this occasion Professor M. V. Subba Rao, who turned 80, was felicitated. We thank the participants of the conference and the contributors to this volume and the referees. Thanks are also due to the administrative staff at the Institute of Mathematical Sciences for their help in organising the conference. Our special thanks are due to Dr. R. Thangadurai and Ms. R. Indra who took upon themselves various organisational matters and followed them up meticulously. Finally we would like to thank Ramanujam Mathematical Society to consider our proceedings as the first volume of the DST-sponsored RMS Lecture Series in Mathematics. Publication of the present volume is co-sponsored by The Institute of Mathematical Sciences, Chennai.

**January 2005**

**S. D. Adhikari**
**R. Balasubramanian**
**K. Srinivas**

# Foreword

I thank Professor Balasubramanian for inviting me to write a foreword to these proceedings of a number theory conference, held at the Institute of Mathematical Sciences, Chennai, during which Professor M. V. Subbarao, who turned eighty, was felicitated.

Professor Subbarao was born on May 4, 1921 in the village Yazali near Bapatla (Guntur District), India. His father was Shri Mathukumalli Narasimha Rao and his mother Venkata Subbamma. He completed his high school education at Guntur and a Master's degree at Presidency College, Madras in 1941, and subsequently entered the department of mathematics of University of Madras to work with Professor Vaidyanathaswami for his doctoral degree.

Later, after working for some years at Presidency College, Madras, he joined the Department of Mathematics, Sri Venkateshwara University, Tirupathi, where he served as professor before moving to the University of Missouri, Columbia (U.S.A.) and later to University of Alberta, Edmonton, Canada in 1963. He spent one year on invitation as Professor and Department Head at Kerala University (1966–67). Professor Subbarao and his wife Mrs. Suseela, have a son, Professor M. Vidyasagar, and a daughter Savithri.

Dr. Subbarao's doctoral thesis is on Functional Analysis and in the beginning he wrote a few papers in Analysis and Topology, however, for most of his mathematical research career he has worked in Number Theory. In particular, Professor Subbarao has contributed many important results and conjectures to the theory of arithmetical functions. He has several works dealing with the distribution of generalized $K$-free integers, partition functions, the Scholz-Brauer problem in addition chains. On the partition function $p(n)$, he had joint papers with George Andrews, D. Bressoud, B. Richmond, L. Carlitz, E. Straus, J. Fabrykowski, A. K. Agarwal et. al., in addition to some forthcoming papers jointly with Imre Katai. In this context Professor Subbarao made the following important conjecture.

*For every positive integer $m$, on every arithmetic progression $r(\text{mod } m)$; $0 \leq r < m-1$, $p(n)$ assumes both even and odd values infinitely often.*

For the cases $m = 2, 3, 4, 5, 6, 8, 10, 12, 16, 20, 40$, the conjecture had been established by the efforts of many mathematicians. In fact, the works of Subbarao, Hirschhorn and Subbarao and Hirschhorn among others like Frank Garvan and D. Stanton established these cases by elegant combinatorial methods.

This conjecture generated a lot of research work. We mention the following result of K. Ono, which is a major achievement in this direction.

For every positive integer $m$, in every arithmetic progression $r(\text{mod } m)$, $0 \leq r < m-1$, $p(n)$ assumes even values infinitely often. Also, if $p(n)$ assumes odd value for a single $n$ in an arithmetical progression, then $p(n)$ assumes odd values for infinitely many $n$ in that arithmetical progression. Quantitative versions of the above result were obtained later by J. P. Serre and S. Ahlgren. The odd case of the conjecture is still open, but has been verified for all $m \leq 10^5$. Professor Subbarao offers a \$500 for a complete proof of his conjecture. He has an analogous conjecture for product partitions.

Many other conjectures and unsolved problems appear in his papers with Erdös, Straus, Katai, Hardy et. al. We just mention one more of his conjectures. If $p_1, \ldots, p_r$ are any distinct primes and $a_1, \ldots, a_r$ positive integers, then $\Pi(p_i^{a_i} - 1)$ divides $((\Pi p_i^{a_i}) - 1)$ only if $r = 1$.

Professor Subbarao collaborated with some forty other researchers, including Paul Erdös, Ernst Straus, L. Carlitz, George Andrews, Imre Katai, Wang Yuan, George Hardy, David Brossoud, M. D. Hirchhorn, K. G. Ramanathan, K. Ramachandra, R. Balasubramanian, R. A. Smith, H. L. Abbott, D. Suryanarayana, R. Sitaramachandra Rao, V. Sivarama Prasad, V. Sitaramaiah, A. Verma, A. K. Agarwal, V. V. S. Sastri, M. Sugunamma et. al. besides his son Vidyasagar. His research publications total around 200.

His work and his pleasant nature brought Professor Subba Rao a number of collaborators. In particular, Professor Subba Rao's Erdös number is one. Indeed, he has eight joint papers with Professor P. Erdös. A number of Indian mathematicians, among them Professor K. G. Ramanathan, Professor R. Balasubramanian and myself, have had the good fortune of working with Professor Subbarao. He has had special interest in Mathematics departments of Sri Venkateshwara and Andhra Universities where his support and encouragement produced excellent mathematicians such as Professors D. Suryanarayana and R. Sita Ramachandra Rao.

**K. Ramachandra**

# Table of Contents

Proc. Int. Conf. – Number Theory No. 1, 2004, pp. 1–6.

# Matrix Theoretic Interpretations of Four Analytic Identities

A. K. Agarwal

Centre for Advanced Study in Mathematics, Panjab University, Chandigarh 160 014, India
e-mail : aka@pu.ac.in

**Abstract.** By using infinite matrices of a special kind, we provide new combinatorial interpretations of four analytic identities.

## 1. Introduction, definitions and the main results

Using infinite matrices we give new combinatorial interpretations of the following identities from [11]

$$\sum_{n=0}^{\infty} \frac{q^{n^2}}{(q,q)_{2n}} = \prod_{n=0}^{\infty} \frac{1}{(1-q^{2n+1})(1-q^{20n+4})(1-q^{20n+16})}, \qquad [11, (79)\text{–}(98)] \tag{1.1}$$

$$\sum_{n=0}^{\infty} \frac{q^{n(2n+1)}}{(q;q)_{2n+1}} = \prod_{n=1}^{\infty} \frac{1}{(1-q^{2n-1})}, \qquad [11, (84)] \tag{1.2}$$

$$\sum_{n=0}^{\infty} \frac{q^{n(n+1)}}{(q;q)_{2n}} = \frac{1}{(q;q)_\infty} \times \prod_{n=1}^{\infty} (1-q^{10n-1})(1-q^{10n-9})(1-q^{20n-8})(1-q^{20n-12})(1-q^{10n}) \quad [11, (99)], \tag{1.3}$$

$$\sum_{n=0}^{\infty} \frac{q^{2n^2}}{(q;q)_{2n}} = \frac{1}{(q;q)_\infty} \times \prod_{n=1}^{\infty} (1-q^{8n-1})(1-q^{8n-7})(1-q^{16n-10})(1-q^{16n-6})(1-q^{8n}), \ [11, (39)\text{–}(83)]. \tag{1.4}$$

In Eqs. (1.1)–(1.4), $(q;q)_n$ is a rising $q$-factorial which in general is defined as follows:

$$(a;q)_n = \prod_{i=0}^{\infty} \frac{(1-aq^i)}{(1-aq^{n+i})}.$$

If $n$ is a positive integer, then obviously

$$(a;q)_n = (1-a)(1-aq)\cdots(1-aq^{n-1}),$$

and

$$(a;q)_\infty = (1-a)(1-aq)(1-aq^2)\cdots.$$

Several combinatorial interpretations of Identities (1.1)–(1.4) are found in the literature. For combinatorial interpretations in terms of classical partitions one can see Gordon [9], Connor [8], Hirschhorn [10], Agarwal and Andrews [4], Agarwal and Padmavathamma [6], Subbarao [12], Subbarao and Agarwal [13]. Other combinatorial interpretations of Identities (1.1)–(1.4) in terms of $n$-colour partitions, lattice paths and generalized Frobenius partitions were given by the author in [1], [2] and [3], respectively. It was pointed out in [5] that the number of $n$-colour partitions of a positive integer $\nu$ equals the number of plane partitions of $\nu$. E. A. Bender and D. E. Knuth [7] proved the following:

**Theorem (Bender and Knuth).** *There is a one-to-one correspondence between plane partitions of $\nu$, on the one hand, and infinite matrices $a_{ij}$ $(i, j \geq 1)$ of non negative integer entries which satisfy,*

$$\sum_{r\geq 1} r\left(\sum_{i+j=r+1} a_{ij}\right) = \nu,$$

*on the other.*

Corresponding to every non negative integer $\nu$ we shall call the matrices of the above theorem $BK_\nu$-matrices (BK for Bender and Knuth). Although, these matrices are infinite but we will represent them by the largest possible square matrices containing at least one non zero entry in the last row (or last column). Thus, for example, we will represent six $BK_3$-matrices by

$$3, \quad \begin{pmatrix}1&0\\1&0\end{pmatrix}, \quad \begin{pmatrix}1&1\\0&0\end{pmatrix} \quad \begin{pmatrix}0&0\\0&1\end{pmatrix}, \quad \begin{pmatrix}0&0&1\\0&0&0\\0&0&0\end{pmatrix}, \quad \begin{pmatrix}0&0&0\\0&0&0\\1&0&0\end{pmatrix}.$$

In this paper we establish a bijection between the $n$-colour partitions of $\nu$ and the $BK_\nu$-matrices. This leads us to new combinatorial interpretations of the Identities (1.1)–(1.4).

We first recall the definitions of $n$-colour partitions and weighted differences from [5].

**Definition (Agarwal and Andrews [5]).** An $n$-colour partition (or a partition with "$n$ copies of $n$") of a positive integer $\nu$ is a partition in which a part of size $n$, $n \geq 1$ can come in $n$-different colours denoted by subscripts $n_1, n_2, \cdots, n_n$.

**Definition (Agarwal and Andrews [5]).** The weighted difference of two parts $m_i$, $n_j$, $m \geq n$ is defined by $m - n - i - j$ and is denoted by $((m_i - n_j))$.

We give here two more definitions.

**Definition 1.** We define a matrix $E_{i,j}$ as an infinite matrix whose $(i, j)$-th entry is 1 and the other entries are all zeros. We call $E_{i,j}$ distinct units of a $BK_\nu$-matrix.

**Definition 2.** The order difference of two units $E_{p,q}$, $E_{r,s} (p + q \geq r + s)$ is defined by $q - s - 2r$ and is denoted by $\{\{E_{p,q} - E_{r,s}\}\}$.

In our next section we shall prove the following theorem:

**Theorem 1.1.** *There is a one-to-one correspondence between the $BK_\nu$-matrices on the one hand, and the n-colour partitions of $\nu$, on the other.*

Using Theorem 1.1 we shall prove the following combinatorial interpretations of Identities (1.1)–(1.4) in Section 3.

**Theorem 1.2.** *Let $A_1(\nu)$ denote the number of $BK_\nu$-matrices such that all entries in the even columns are zero and the order difference between any two units $E_{p,q}$ and $E_{r,s}$ with $p + q \geq r + s$ is non negative. Let $B_1(\nu)$ denote the number of partitions of $\nu$ into parts $\not\equiv 0, \pm 2, \pm 6, \pm 8, 10 \pmod{20}$. Then $A_1(\nu) = B_1(\nu)$.*

**Example.** $A_1(5) = 4$, since the relevant $BK_5$-matrices are

$$\begin{pmatrix} 0&0&0&0&1\\ 0&0&0&0&0\\ 0&0&0&0&0\\ 0&0&0&0&0 \end{pmatrix}, \quad \begin{pmatrix} 0&0&0\\ 0&0&0\\ 0&0&1 \end{pmatrix}, \quad \begin{pmatrix} 0&0&0&0&0\\ 0&0&0&0&0\\ 0&0&0&0&0\\ 0&0&0&0&0\\ 1&0&0&0&0 \end{pmatrix}, \quad \begin{pmatrix} 1&0&0\\ 0&0&1\\ 0&0&0 \end{pmatrix}.$$

Also, $B_1(5) = 4$, since the relevant partitions are $5, 41, 31^2, 1^5$.

**Theorem 1.3.** *Let $A_2(\nu)$ denote the number of $BK_\nu$-matrices such that all entries in even columns are zero and the order difference of any two units is $\geq 2$. Let $B_2(\nu)$ denote the number of partitions of $\nu$ into distinct parts. Then $A_2(\nu) = B_2(\nu)$.*

**Theorem 1.4.** *Let $A_3(\nu)$ denote the number of $BK_\nu$-matrices such that all entries in the first row as well as in all even columns are zero and the order difference of any two units $\geq -2$. Let $B_3(\nu)$ denote the number of partitions of $\nu$ into parts $\not\equiv 0, \pm 1, \pm 8, \pm 9, 10$ $\pmod{20}$. Then $A_3(\nu) = B_3(\nu)$.*

**Theorem 1.5.** *Let $A_4(\nu)$ denote the number of $BK_\nu$-matrices in which all entries in the first row as well as in all even columns are zero and the order difference of any two units is non negative. Let $B_4(\nu)$ denote the number of partitions of $\nu$ into parts $\not\equiv 0, \pm 1, \pm 6, \pm 7, 8 \pmod{16}$. Then $A_4(\nu) = B_4(\nu)$.*

In the proofs of Theorems 1.2–1.5 we shall make use of the following results from [1]:

**Theorem 1.6 (Theorem 1.1, [1]).** *Let $C_1(\nu)$ denote the number of n-colour partitions of $\nu$ wherein each pair of parts has non negative weighted difference and even parts appear with even subscripts and odd with odd subscripts. Then $C_1(\nu) = B_1(\nu)$.*

**Theorem 1.7 (Theorem 1.4, [1]).** *Let $C_2(\nu)$ denote the number of n-colour partitions of n wherein each pair of parts has weighted difference $\geq 1$ and even parts appear with even subscripts and odd with odd. Then $C_2(\nu) = B_2(\nu)$.*

**Theorem 1.8 (Theorem 1.5, [1]).** *Let $C_3(\nu)$ denote the number of n-colour partitions of $\nu$ where in each pair of parts has weighted difference $\geq -2$ and even parts appear with even subscripts and the odd parts with odd subscripts $> 1$. Then $C_3(\nu) = B_3(\nu)$.*

**Theorem 1.9 (Theorem 1.6, [1]).** *Let $C_4(\nu)$ denote the number of n-colour partitions of $\nu$ such that each pair of parts has non negative weighted difference and the even parts appear with even subscripts and odd with odd subscripts $> 1$. Then $C_4(\nu) = B_4(\nu)$.*

## 2. Proof of theorem 1.1

Let $\Delta = a_{1,1}E_{1,1}+a_{1,2}E_{1,2}+\cdots+a_{2,1}E_{2,1}+a_{2,2}E_{2,2}+\cdots+a_{3,1}E_{3,1}+a_{3,2}E_{3,2}+\cdots$ be a $BK_\nu$-matrix, where $a_{i,j}$ are non negative integers which denote the multiplicities of $E_{i,j}$. We map each part $E_{p,q}$ of $\Delta$ to a single part $m_i$ of an $n$-colour partition of $\nu$. The mapping $f$ is

$$f : E_{p,q} \longrightarrow (p+q-1)_p,$$

and the inverse mapping $f^{-1}$ is easily seen to be

$$f^{-1} : m_i \longrightarrow E_{i,m-i+1}.$$

Under this mapping we see that each $BK_\nu$—matrix uniquely corresponds to an $n$-colour partition of $\nu$ and vice-versa. □

To illustrate this bijection we consider the case for $\nu = 3$. (Table 1)

**Table 1.** .

| $K_3$—*matrices* $\Delta$ | $f(\Delta)$ |
|---|---|
| $(3) \quad = 3E_{1,1}$ | $31_1 = 1_1 + 1_1 + 1_1$ |
| $\begin{pmatrix} 1 & 1 \\ 0 & 0 \end{pmatrix} = E_{1,1} + E_{1,2}$ | $1_1 + 2_1$ |
| $\begin{pmatrix} 1 & 0 \\ 1 & 0 \end{pmatrix} = E_{1,1} + E_{2,1}$ | $1_1 + 2_2$ |
| $\begin{pmatrix} 0 & 0 & 1 \\ 0 & 0 & 0 \\ 0 & 0 & 0 \end{pmatrix} = E_{1,3}$ | $3_1$ |
| $\begin{pmatrix} 0 & 0 \\ 0 & 1 \end{pmatrix} = E_{2,2}$ | $3_2$ |
| $\begin{pmatrix} 0 & 0 & 0 \\ 0 & 0 & 0 \\ 1 & 0 & 0 \end{pmatrix} = E_{3,1}$ | $3_3$ |

## 3. Proofs of theorems 1.2–1.5

In this section we give detail proof of Theorem 1.2 and hints for proofs of the remaining theorems. The interested reader can supply the details or obtain them from the author.

For proving Theorem 1.2 we use Theorem 1.6. In view of Theorem 1.1 we have only to prove that if a $BK_\nu$-matrix $\Delta$ is enumerated by $A_1(\nu)$, then in the $n$-colour partition $f(\Delta)$ the even parts appear with even subscripts and the odd with odd subscripts and each pair of parts has non negative weighted difference, and conversely, if an $n$-colour partition $\Pi$ is enumerated by $C_1(\nu)$ then in the $BK_\nu$-matrix $f^{-1}(\Pi)$ all entries in the even columns are zero and the order difference of any two units is non negative. Let $E_{p,q}, E_{r,s}(p+q \geq r+s)$ be two units of a $BK_\nu$-matrix $\Delta$ enumerated by $A_1(\nu)$ which correspond to two $n$-colour parts $m_i, n_j$ of $f(\Delta)$. Then $m_i = (p+q-1)_p$ and $n_j = (r+s-1)_r$.

Since $p+q \geq r+s$, therefore $m \geq n$ and

$$\begin{aligned}((m_i - n_j)) &= (p+q-1) - p - (r+s-1) - r = q - s - 2r \\ &= \{\{E_{p,q} - E_{r,s}\}\} \geq 0.\end{aligned}$$

Furthermore, since in $E_{p,q}$, $q$ is odd, we see that $p+q-1$ and $p$ have the same parity which implies that in $f(\Delta)$ even parts appear with even subscripts and odd with odd. Thus $f(\Delta)$ is enumerated by $C_1(\nu)$.

Next, suppose $\Pi$ is an $n$-colour partition enumerated by $C_1(\nu)$ and $m_i, n_j (m \geq n)$ are its two parts which correspond to two units $E_{p,q}$ and $E_{r,s}$ of $f^{-1}(\Pi)$. Then $E_{p,q} = E_{i,m-i+1}$ and $E_{r,s} = E_{j,n-j+1}$.

Since $m \geq n$, we have $p+q = m+1 \geq n+1 = r+s$, and

$$\begin{aligned}\{\{E_{p,q} - E_{r,s}\}\} &= \{\{E_{i,m-i+1} - E_{j,n-j+1}\}\} \\ &= (m-i+1) - (n-j+1) - 2j \\ &= m - n - i - j \\ &= ((m_i - n_j)) \geq 0.\end{aligned}$$

Furthermore, since in $\Pi$ even parts appear with even subscripts and odd with odd, we see that $m-i+1$ and $n-j+1$ are both odd. This implies that in $f^{-1}(\Pi)$ all entries in even columns are zero. Thus $f^{-1}(\Pi)$ is enumerated by $A_1(\nu)$. This completes the proof of Theorem 1.2. □

In the proofs of Theorems 1.3–1.5 we use Theorems 1.7–1.9, respectively, and the fact that the condition' in $m_i, i > 1$' for an $n$-colour partition $\Pi$ is equivalent to saying that' in $f^{-1}(\Pi)$, the first row does not contain a non zero element and vice-versa.

## References


[1] A. K. Agarwal, Rogers-Ramanujan identities for $n$-color partitions, *J. Number Theory*, **28** (1988) 299–305

[2] A. K. Agarwal, Lattice paths and $n$-color partitions, *Utilitas Math*, **53** (1998) 71–80

[3] A. K. Agarwal, Identities and generating functions for certain classes of $F$-partitions, *ARS Combinatoria*, **57** (2000) 65–75

[4] A. K. Agarwal and G. E. Andrews, Hook differences and lattice paths, *Journal of statistical Planning and Inference*, **V. 14** (1986) 5–14

[5] A. K. Agarwal and G. E. Andrews, Rogers-Ramanujan identities for partitions with "$N$ copies of $N$", *J. Combin. Theory, Ser. A.*, **V. 45(1)** (1987) 40–49

[6] A. K. Agarwal and Padmavathamma, Rogers-Ramanujan type identities for Burge's restricted partition pairs via restricted Frobenius partitions, *Number theory and discrete mathematics* (Chandigarh, 2000), *Trends Math.*, Birkhäuser, Basel, (2002) 53–60

[7] E. A. Bender and D. E. Knuth, Enumeration of plane partitions, *J. Combin. Theory (A)*, **13** (1972) 40–54

[8] W. G. Connor, Partition theorems related to some identities of Rogers and Watson, *Trans. Amer. Math. Soc.* **214** (1975) 95–111

[9] B. Gordon, Some continued fractions of the Rogers-Ramanujan type, *Duke Math. J.* **32** (1965) 741–748

[10] M. D. Hirschhorn, Some partition theorems of the Rogers-Ramanujan type, *J. Combin. Theory Ser A*, **27(1)** (1979) 33–37

[11] L. J. Slater, Further identities of the Rogers-Ramanujan type, *Proc. London Math. Soc.* **54** (1951–52) 147–167

[12] M. V. Subbarao, Some Rogers-Ramanujan type partition theorems, Pacific *J. Math.* **120** (1985) 431–435

[13] M. V. Subbarao and A. K. Agarwal, Further theorems of the Rogers-Ramanujan type theorems, *Canad. Math. Bull.* **31(2)** (1988) 210–214

Proc. Int. Conf. – Number Theory No. 1, 2004, pp. 7–14.

# Taylor–Wiles Systems

Eknath Ghate

School of Mathematics, Tata Institute of Fundamental Research, Homi Bhabha Road, Mumbai 400 005, India
e-mail : eghate@math.tifr.res.in

June 17, 2004

## 1. Introduction

The purpose of this note is to recall the statements and proofs of some theorems in commutative algebra originating in the work of Wiles [8] and Taylor–Wiles [7]. In particular we wish to recall a useful freeness result for certain modules over complete local noetherian algebras over $p$-adic rings.

The key commutative algebra result of Taylor–Wiles [7] is an isomorphism criterion for a morphism between complete local noetherian algebras. In fact the above mentioned freeness result is used as a hypothesis in their work. However later refinements due to Faltings, Diamond [1] and Fujiwara [2] have shown that it is possible to deduce the freeness result as a byproduct of rather than an ingredient in the isomorphism criterion. It is this refined version of the Taylor–Wiles machine, in particular the approach of Fujiwara in [2], that we shall expose here.

Our interest in the freeness result is because of its relevance in the study of congruence primes for Hilbert modular forms. In [3] we use it to study the freeness of suitably localized middle dimensional integral cohomology groups of Hilbert Blumenthal varieties considered as modules over localized Hecke algebras. We use this freeness to show how the (ordinary) primes of congruence for a Hilbert modular form are captured by the value of its adjoint $L$-function.

Various expositions of the results described in this note are already in the literature. Apart from the original sources mentioned above see, for instance, the book of Hida [4] and the work of Skinner and Wiles [6]. The exposition here is primarily based on our own efforts to understand the work of the above authors. We wish to thank S. Bhatwadekar, A. Brown, and H. Hida for some helpful conversations.

## 2. Taylor–Wiles systems

Let $F$ be a number field. Let $\mathcal{O}_F$ denote the ring of integers of $F$. We fix an odd prime $p$, and an extension $\mathcal{O}$ of $\mathbb{Z}_p$ that is the ring of integers in a finite extension of $\mathbb{Q}_p$. Consider the set $\mathcal{Q}$ defined via

$$\mathcal{Q} = \{\mathfrak{q} \subset \mathcal{O}_F | \mathfrak{q} \text{ prime}, \ \mathrm{N}_{F|\mathbb{Q}}(\mathfrak{q}) \equiv 1 \ (\mathrm{mod}\ p)\}. \quad (1)$$

For each finite subset $Q \subset \mathcal{Q}$, let

- $\Delta_Q =$ the $p$-Sylow subgroup of $\prod_{\mathfrak{q} \in Q} (\mathcal{O}_F/\mathfrak{q})^\times$,
- $\mathcal{O}[\Delta_Q] =$ the corresponding group algebra over $\mathcal{O}$, and,
- $\mathfrak{a}_Q =$ the augmentation ideal of $\mathcal{O}[\Delta_Q]$.

**Definition 1 (Taylor–Wiles system).** For each $m = 1, 2, \ldots$, fix a finite subset $Q_m \subset \mathcal{Q}$. A Taylor–Wiles system consists of a complete, noetherian, local $\mathcal{O}$-algebra $R$, and a collection of triples $(R_m, T_m, M_m)$, indexed by $m = 1, 2, \ldots$, where

- $R_m$ is a complete, noetherian, local $\mathcal{O}$-algebra, with a map $R_m \twoheadrightarrow R$,
- $T_m$ is a complete, noetherian, local $\mathcal{O}$-algebra, with a map $R_m \twoheadrightarrow T_m$,
- $M_m$ is a $T_m$-module,

satisfying:

- $R_m$ (so also $T_m$) is an $\mathcal{O}[\Delta_{Q_m}]$-algebra, and,
- $R_m/\mathfrak{a}_{Q_m} R_m \xrightarrow{\sim} R$.

## 3. Main results

**Theorem 1 (Freeness Criterion).** *Let $R$, $(R_m, T_m, M_m)$ be a Taylor–Wiles system. Assume that the following conditions hold:*

(TW1) *If $\mathfrak{q} \in Q_m$ then $\mathrm{N}_{F|\mathbb{Q}}(\mathfrak{q}) \equiv 1 (\mathrm{mod}\ p^m)$,*
(TW2) *The cardinality of $Q_m$ is independent of $m$, say $|Q_m| = r$ for all $m$,*
(TW3) *$R_m$ is generated by at most $r$ elements as an $\mathcal{O}$-algebra,*
(TW4) *As an $\mathcal{O}[\Delta_{Q_m}]$-module, $M_m$ is free of finite rank $s > 0$ independent of $m$,*
(TW5) *$\exists$ $R$-module $M : M_m/\mathfrak{a}_{Q_m} M_m = M$, $\forall m$, as $R$-modules.*

*Then $M$ is a free $R$-module.*

Note that a Taylor–Wiles system yields a collection of diagrams

$$\begin{array}{ccc} R_m & \twoheadrightarrow & T_m \\ \downarrow & & \downarrow \\ R & & \mathrm{End}(M_m), \end{array}$$

which modulo the augmentation ideals of the $\mathcal{O}[\Delta_{Q_m}]$ induce the diagrams

$$\begin{array}{ccc} R_m/\mathfrak{a}_{Q_m} R_m & \twoheadrightarrow & T_m/\mathfrak{a}_{Q_m} T_m \\ \| & & \downarrow \\ R & \xrightarrow{\pi_m} & \mathrm{End}(M_m/\mathfrak{a}_{Q_m} M_m), \end{array}$$

where $\pi_m$ is defined so that the above diagram commutes. (TW5) implies that the kernel of $\pi_m : R \to \mathrm{End}_{\mathcal{O}}(M_m/\mathfrak{a}_{Q_m} M_m)$ is independent of $m$. Let $T = R/\ker(\pi_m)$ for any $m$.

**Theorem 2 (Isomorphism criterion).** *Under* (TW1-5) *$R$ is a complete intersection ring which is free of finite rank over $\mathcal{O}$. Moreover, the natural map*

$$R \twoheadrightarrow T$$

*is an isomorphism. In particular $M$ is a free $T$-module.*

*Proof.* We prove Theorems 1 and 2 simultaneously. The proof is essentially a recall of the approach of Fujiwara in [2]. Fix $m \in \mathbb{N}$. For each $\mathfrak{q} \in Q_m$, let $\Delta_{\mathfrak{q}}$, be the $p$-Sylow subgroup of $(\mathcal{O}_F/\mathfrak{q})^\times$. By (TW1) we know that this is cyclic of order at least $p^m$. Let $\delta_{\mathfrak{q}}$ be a generator of $\Delta_{\mathfrak{q}}$. Suppose that $1 \leq n \leq m$. Note that $\Delta_{\mathfrak{q}}/\langle \delta_{\mathfrak{q}}^{p^n} \rangle$ is cyclic of order $p^n$. Thus,

$$\frac{\mathcal{O}[\Delta_{\mathfrak{q}}]}{\langle \delta_{\mathfrak{q}}^{p^n} - 1 \rangle} \xrightarrow{\sim} \frac{\mathcal{O}[[S]]}{\langle (1+S)^{p^n} - 1 \rangle},$$

via $\delta_{\mathfrak{q}} \mapsto 1 + S$.

By (TW2) we know that the cardinality of $Q_m$ is $r$ (and this number is independent of $m$). Suppose that $Q_m = \{\mathfrak{q}_1, \ldots, \mathfrak{q}_r\}$. Let $I_n \subset \mathcal{O}[\Delta_{Q_m}]$ be the ideal generated by $p^n, \delta_{\mathfrak{q}_1}^{p^n} - 1, \ldots, \delta_{\mathfrak{q}_r}^{p^n} - 1$. Then we have

$$\frac{\mathcal{O}[\Delta_{Q_m}]}{I_n} \xrightarrow{\sim} \frac{\mathcal{O}[[S_1, \ldots, S_r]]}{\langle p^n, (1+S_1)^{p^n} - 1, \ldots, (1+S_r)^{p^n} - 1 \rangle},$$

via $\delta_{\mathfrak{q}_i} \mapsto 1 + S_i$, for $i = 1, \ldots, r$.

Let us set $A_n = \mathcal{O}[\Delta_{Q_m}]/I_n$. It is an easy computation to check that $|A_n| = p^{ntp^{nr}}$, where $t = \mathrm{rank}_{\mathbb{Z}_p}(\mathcal{O})$. In particular the cardinality of $A_n$ is independent of $m$. Now define

$$R_{n,m} = \mathrm{image}\left(\frac{R_m}{I_n R_m} \to \mathrm{End}_{A_n}\left(\frac{M_m}{I_n M_m}\right)\right),$$

$$\widetilde{R}_{n,m} = \frac{R_{n,m}}{\mathfrak{a}_{Q_m} R_{n,m}}.$$

By (TW 4) $M_m$ is free over $\mathcal{O}[\Delta_{Q_m}]$ of rank $s$ (independent of $m$). Thus there are natural maps

$$A_n \xrightarrow{\alpha} R_{n,m} \overset{\beta}{\hookrightarrow} \mathrm{End}_{A_n}\left(\frac{M_m}{I_n M_m}\right) = \mathrm{M}_s(A_n).$$

Since $\beta$ is injective we see that $|R_{n,m}| \leq |A_n|^{s^2}$ is bounded independently of $m$. Since $\beta \circ \alpha$ is the natural map making $M_s(A_n)$ an $A_n$-module, it is injective as well. This implies that $\alpha$ is also injective. This will be used below.

By (TW3), $R_m$ is generated by $r$ elements over $\mathcal{O}$, and thus, so is $R_{n,m}$. Pick generators $f_1, \ldots, f_r$ of $R_{n,m}$.

Now fix $n$, and suppose $m, m' \geq n$. We say two triples

$$\big((R_{n,m}, \alpha, \beta), \widetilde{R}_{n,m}, (f_1, \ldots, f_r)\big) \;\&\; \big((R_{n,m'}, \alpha', \beta'), \widetilde{R}_{n,m'}, (f_1', \ldots, f_r')\big)$$

are isomorphic if there is an isomorphism of $A_n$-algebras

$$\iota : R_{n,m} \xrightarrow{\sim} R_{n,m'}$$

(inducing $\iota : \widetilde{R}_{n,m} \xrightarrow{\sim} \widetilde{R}_{n,m'}$), such that $\iota(f_j) = f'_j$ and such that the following diagram commutes:

$$\begin{array}{ccccc} A_n & \xhookrightarrow{\alpha} & R_{n,m} & \xhookrightarrow{\beta} & \mathrm{M}_s(A_n) \\ \| & & \downarrow \iota & & \| \\ A_n & \xhookrightarrow{\alpha} & R_{n,m'} & \xhookrightarrow{\beta'} & \mathrm{M}_s(A_n). \end{array}$$

The induced diagram

$$\begin{array}{ccc} R_{n,m} & \twoheadrightarrow & \widetilde{R}_{n,m} \\ \downarrow \iota & & \downarrow \iota \\ R_{n,m'} & \twoheadrightarrow & \widetilde{R}_{n,m'} \end{array}$$

is automatically commutative. Since the cardinality of $R_{n,m}$ is bounded independently of $m$, we see that for $n = 1$, there exists infinitely $m \geq 1$ (call this set $\mathbb{N}_1 \subset \mathbb{N}$) such that the

$$\left((R_{1,m}, \alpha, \beta), \widetilde{R}_{1,m}, (f_1, \ldots, f_r)\right),$$

are mutually isomorphic.

Suppose, for purposes of induction, that we have constructed $\mathbb{N}_n \subset \mathbb{N}_{n-1} \subset \cdots \subset \mathbb{N}_1 \subset \mathbb{N}$ such that

$$\left((R_{n,m}, \alpha, \beta), \widetilde{R}_{n,m}, (f_1, \ldots, f_r)\right),$$

are mutually isomorphic for all $n \in \mathbb{N}_n$. Then, again, by varying $m \geq n+1$ in $\mathbb{N}_n$ we may pick an infinite subset $\mathbb{N}_{n+1} \subset \mathbb{N}_n$, such that

$$\left((R_{n+1,m}, \alpha, \beta), \widetilde{R}_{n+1,m}, (f_1, \ldots, f_r)\right),$$

are mutually isomorphic for $m \in \mathbb{N}_{n+1}$.

Now, for each $n$, let $m(n)$ denote the minimal element in $\mathbb{N}_n$, and let the corresponding triple be

$$\left((R_{n,m(n)}, \alpha, \beta), \widetilde{R}_{n,m(n)}, (f_1^{(n)}, \ldots, f_r^{(n)})\right).$$

The following diagram commutes

$$\begin{array}{ccccc} \dfrac{R_{m(n+1)}}{I_{n+1}R_{m(n+1)}} & \twoheadrightarrow & R_{n+1,m(n+1)} & \hookrightarrow & \mathrm{End}_{A_{n+1}}\left(\dfrac{M_{m(n+1)}}{I_{n+1}M_{m(n+1)}}\right) \\ \downarrow \bmod I_n & & \downarrow & & \\ \dfrac{R_{m(n+1)}}{I_nR_{m(n+1)}} & \twoheadrightarrow & R_{n,m(n+1)} & \hookrightarrow & \mathrm{End}_{A_n}\left(\dfrac{M_{m(n+1)}}{I_nM_{m(n+1)}}\right). \end{array}$$

Moreover $R_{n,m(n+1)}$ (and its tuple) is isomorphic to $R_{n,m(n)}$ (and its tuple) since both $m(n+1)$ and $m(n) \in \mathbb{N}_n$. Thus there is a surjective map $R_{n+1,m(n+1)} \twoheadrightarrow R_{n,m(n)}$ obtained by going modulo $I_n$. Note that under this map $(f_1^{(n+1)}, \ldots, f_r^{(n+1)}) \mapsto (f_1^{(n)}, \ldots, f_r^{(n)})$. We also obtain an induced surjection $\widetilde{R}_{n+1,m(n+1)} \twoheadrightarrow \widetilde{R}_{n,m(n)}$ as well. The upshot is that the transition maps obtained by going modulo $I_n$ yield a projective system of tuples

$$\Big((R_{n,m(n)}, \alpha, \beta), \widetilde{R}_{n,m(n)}, (f_1^{(n)}, \ldots, f_r^{(n)})\Big)$$

as $n$ varies. Denote the projective limit by

$$\Big((R_\infty, \alpha_\infty, \beta_\infty), \widetilde{R}_\infty, (f_1^{(\infty)}, \ldots, f_r^{(\infty)})\Big).$$

Let us now set

$$A := \varprojlim A_n = \mathcal{O}[[S_1, \ldots S_r]], \text{ and } B := \mathcal{O}[[T_1, \ldots T_r]].$$

By the injectivity of the $\alpha$ and $\beta$ at each level, we have injective maps

$$A \overset{\alpha_\infty}{\hookrightarrow} R_\infty \overset{\beta_\infty}{\hookrightarrow} \mathrm{M}_s(A). \tag{2}$$

Consider the map

$$\begin{aligned} B &\twoheadrightarrow R_\infty \\ T_i &\mapsto f_i^{(\infty)}. \end{aligned} \tag{3}$$

The maps (2) show that $R_\infty$ is a torsion free $A$-module of finite type, and so by the going-up and going-down theorems, the Krull dimension of $R_\infty$ is given by $\dim(R_\infty) = r+1$. This shows that the map (3) must be injective. For if it were not, we could always choose some $0 \neq f$ in its kernel $K$, and a height 1 prime $P$ of $B$ containing $f$, and we would get the absurdity:

$$r+1 = \dim R_\infty = \dim B/K \leq \dim B/(f) = \dim B/P = r.$$

This shows that $R_\infty$ is isomorphic to the power series ring $B = \mathcal{O}[[T_1, \ldots T_r]]$.

We now claim that $\widetilde{R}_\infty$ is a complete intersection, free of finite rank over $\mathcal{O}$. To see this note that we have an exact sequence

$$R_\infty^r \longrightarrow R_\infty \longrightarrow \widetilde{R}_\infty \longrightarrow 0$$

where the first map is given by $(x_i) \mapsto \sum_{i=1}^r S_i x_i$. This shows that $\widetilde{R}_\infty$ is the homomorphic image of a power series ring $B$ over $\mathcal{O}$ in $r$ variables, with kernel generated by $r$ elements which are part of a $B$-regular sequence, hence $\widetilde{R}_\infty$ is a complete intersection. The fact that $\widetilde{R}_\infty$ is free of finite rank over $\mathcal{O}$ follows from the following lemma:

**Lemma 1.** *Let $A$ and $B$ be power series rings over $\mathcal{O}$ in the same number of variables. Suppose that there is an injective local ring homomorphism $A \hookrightarrow B$. Assume that $B$ is of finite type as an $A$-module. Then*

i) *$B$ is $A$-free.*
ii) *If $L$ is a $B$-module of finite type, then $L$ is $A$-free $\Longleftrightarrow$ $L$ is $B$-free.*

*Proof.* Note that ii) $\Rightarrow$ i), but we shall prove i) first and then use it to deduce ii).

Under the hypotheses of the Lemma we have

$$B \text{ is } A\text{-free} \Longleftrightarrow B \text{ is } A\text{-projective} \Longleftrightarrow B \text{ is } A\text{-flat.}$$

The first equivalence follows since $A$ is a local ring. The second equivalence follows since $B$ is a finitely presented $A$-module (indeed $B$ is of finite type over $A$, so there is a map of $A$-modules $0 \to K \to A^n \to B \to 0$, and since $A$ is noetherian, $K$ is also finitely generated as an $A$-module).

Let $\mathfrak{m}$ be the maximal ideal of $A$, and $\mathfrak{n}$ the maximal ideal of $B$. We claim that $\sqrt{\mathfrak{m}B} = \mathfrak{n}$. Indeed let $\mathfrak{p}$ be a prime ideal of $B$ containing $\mathfrak{m}B$. Then $\mathfrak{p} \cap A = \mathfrak{m}$ since $\mathfrak{m}$ is maximal. Thus we have an induced inclusion $A/\mathfrak{m} \hookrightarrow B/\mathfrak{p}$ which is finite and therefore integral. In particular $\dim(B/\mathfrak{p}) = \dim(A/\mathfrak{m}) = 0$ and so $\mathfrak{p} = \mathfrak{n}$ is a maximal ideal proving our claim. This shows that $\dim(B/\mathfrak{m}B) = 0$. Now we note that the regular local ring $B$ is Cohen-Macaulay so that Theorem 23.1 of [5] applies to show that $B$ is $A$ flat, proving i).

We now give an alternative proof of i). The following well known formula due to Auslander-Buchsbaum (cf. [5], Theorem 19.1) is useful.

(AB) Say that $S$ is a noetherian local ring, and $N$ is a finite $S$-module. If $\operatorname{projdim}_S(N) < \infty$, then $\operatorname{depth}_S(N) + \operatorname{projdim}_S(N) = \dim S$.

By the Auslander-Buchsbaum formula we have $\operatorname{depth}_A(B) \leq \dim(A)$. On the other hand, since $A$ is a regular local ring, the maximal ideal $\mathfrak{m}_A = (x_1, \dots, x_d)$ of $A$ is generated by $d$ elements, where $d = \dim(A)$. We claim that $x_1, \dots, x_d$ is a $B$-regular sequence as well. To see this note that since the radical of $\mathfrak{m}B$ is $\mathfrak{n}$, there is a positive integer $\nu$ such that $\mathfrak{n}^\nu \subset \mathfrak{m}B \subset \mathfrak{n}$. In particular $\mathfrak{m}B$ is an $\mathfrak{n}$-primary ideal, and so by definition $x_1, \dots, x_d$ is a system of parameters for $B$. But, as already mentioned above, $B$ is Cohen-Macaulay, so by Theorem 17.4(iii) of [5], $x_1, \dots, x_d$ is a $B$-regular sequence as claimed. Consequently $\operatorname{depth}_A(B) \geq d = \dim(A)$. Thus $\operatorname{depth}_A(B)$ is in fact equal to $\dim(A)$. By (AB) we have $\operatorname{projdim}_A(B) = 0$. This implies $B$ is $A$-projective proving i).

Let us now turn to ii). Suppose that $L$ is a $B$-module of finite type. We prove by induction on $\operatorname{projdim}_B(L)$ that $\operatorname{projdim}_B(L) = \operatorname{projdim}_A(L)$. If $\operatorname{projdim}_B(L) = 0$, that is if $L$ is $B$-free, then i) implies that $L$ is $A$-free and $\operatorname{projdim}_A(L) = 0$. So suppose that $\operatorname{projdim}_B(L) > 0$. Pick an exact sequence of $B$-modules

$$0 \to K \to B^m \to L \to 0.$$

Note that $\operatorname{projdim}_B(L) = \operatorname{projdim}_B(K) + 1$. On the other hand the above exact sequence may be considered as an exact sequence of $A$-modules as well. So $\operatorname{projdim}_A(L) = \operatorname{projdim}_A(K) + 1$. By the induction hypothesis $\operatorname{projdim}_B(K) = \operatorname{projdim}_A(K)$. So $\operatorname{projdim}_B(L) = \operatorname{projdim}_A(L)$. In particular ii) follows. $\square$

Returning to our situation, since $A \hookrightarrow B \hookrightarrow \mathrm{M}_s(A)$, $B$ is indeed an $A$-module of finite type. By Lemma 1, $B = R_\infty$ is $A$-free (of finite rank), and so $\widetilde{R}_\infty$ is $\mathcal{O}$-free (of the same rank).

We now claim that $\widetilde{R}_\infty = T$. Let

$$L = \varprojlim_n M_{m(n)}/I_n M_{m(n)} = \varprojlim A_n^s = A^s.$$

Since $L$ is a finite $A$-module it is fact a finite $B$ - module. Lemma 1 implies that $L$ is a free $B$-module (of finite rank), and so $L/I_nL = M_{m(n)}/I_nM_{m(n)}$ is $R_\infty/I_nR_\infty = R_{n,m(n)}$-free. The following general principle is now useful: if $X$ is a free of finite rank $C$-module and $\mathfrak{a} \subset C$ is an ideal, then $C/\mathfrak{a}$ acts *faithfully* on $X/\mathfrak{a}X$, and so we may consider $C/\mathfrak{a} \subset \mathrm{End}(X/\mathfrak{a}X)$. Applying this to our situation we see

$$\widetilde{R}_{n,m(n)} \subset \mathrm{End}_{\mathcal{O}}\left(M_{m(n)}/(I_n + \mathfrak{a}_{Q_{m(n)}})M_{m(n)}\right). \tag{4}$$

By (TW5) $\ker \pi_{m(n)}$ is independent of $m$. Let $T = R/\ker \pi_{m(n)}$. Actually we shall identify $T$ with its image under $\pi_{m(n)}$ in what follows. Let $S_{n,m(n)}$ be the image of $R_{m(n)}$ in $\mathrm{End}_{\mathcal{O}}(M_{m(n)}/(I_n + \mathfrak{a}_{Q_{m(n)}})M_{m(n)})$. Then there is a natural map $g_n : T \twoheadrightarrow S_{n,m(n)}$. But by (4) $S_{n,m(n)} = \widetilde{R}_{n,m(n)}$. Thus we obtain the following commutative diagram:

$$\begin{array}{ccccc}
R_{m(n)} & \twoheadrightarrow & T & \hookrightarrow & \mathrm{End}_{\mathcal{O}}\left(M_{m(n)}/\mathfrak{a}_{Q_{m(n)}}M_{m(n)}\right) \\
\downarrow & & \downarrow g_n & & \\
R_{n,m(n)} & \twoheadrightarrow & \widetilde{R}_{n,m(n)} & \hookrightarrow & \mathrm{End}_{\mathcal{O}}\left(M_{m(n)}/(I_n + \mathfrak{a}_{Q_{m(n)}})M_{m(n)}\right).
\end{array}$$

Note that the map $g_n$ is the restriction to $T$ of the natural map $\mathrm{M}_s(\mathcal{O}) \twoheadrightarrow \mathrm{M}_s(\mathcal{O}/p^n)$. As such $\ker(g_n) = T \cap p^n\mathrm{M}_s(\mathcal{O})$. Now we have

$$T = \varprojlim T/p^nT = \varprojlim \frac{T}{T \cap p^n\mathrm{M}_s(\mathcal{O})} = \varprojlim \widetilde{R}_{n,m(n)} = \widetilde{R}_\infty,$$

where the second equality follows from the fact that the collection of ideals $p^nT$ and $T \cap p^n\mathrm{M}_s(\mathcal{O})$ of $T$ define the same topology on $T$ (by the Artin-Rees lemma, cf. [5], Theorem 8.5).

We now show that $R \xrightarrow{\sim} \widetilde{R}_\infty$. Let

$$K_n = \ker(R_\infty \twoheadrightarrow R_{n,m(n)}),$$

and let $\mathfrak{m}_\infty$ be the maximal ideal of $R_\infty$. Then for each $N > 0$, there exists $n(N) \gg 0$, such that $K_{n(N)} \subset \mathfrak{m}_\infty^N$. Let $\mathfrak{m}_{R_{m(n)}}$ respectively $\mathfrak{m}_{R_{n,m(n)}}$ be the maximal ideal of $R_{m(n)}$ respectively $R_{n,m(n)}$. We claim that for the choice of $n(N)$ made above the natural map

$$R_{m(n(N))}/\mathfrak{m}^N_{R_{m(n(N))}} \twoheadrightarrow R_{n(N),m(n(N))}/\mathfrak{m}^N_{R_{n(N),m(n(N))}} \tag{5}$$

is an isomorphism. It suffices to construct a surjection in the reverse direction. By our choice of $n(N)$ we have $R_\infty/\mathfrak{m}_\infty^N \xrightarrow{\sim} R_{n(N),m(n(N))}/\mathfrak{m}^N_{R_{n(N),m(n(N))}}$. On the other hand, by (TW3), $R_{m(n(N))}$ is generated by $r$ elements as an $\mathcal{O}$-algebra. This means that there is a surjective map $R_\infty \to R_{m(n(N))}$, which in turn induces a surjective map $R_\infty/\mathfrak{m}_\infty^N \twoheadrightarrow R_{m(n(N))}/\mathfrak{m}_{R_{m(n(N))}}^N$. Thus we have constructed a surjection

$$R_{n(N),m(n(N))}/\mathfrak{m}_{R_{n(N),m(n(N))}}^N \twoheadrightarrow R_{m(n(N))}/\mathfrak{m}_{R_{m(n(N))}}^N,$$

proving that (5) is an isomorphism.

Now consider the following commutative diagram with exact rows:

$$\begin{array}{ccccccc} R_{m(n(N))}^r & \longrightarrow & R_{m(n(N))} & \longrightarrow & R & \longrightarrow & 0 \\ \downarrow & & \downarrow & & \downarrow & & \\ R_{n(N),m(n(N))}^r & \longrightarrow & R_{n(N),m(n(N))} & \longrightarrow & \widetilde{R}_{n(N),m(n(N))} & \longrightarrow & 0. \end{array}$$

Let $C_{n(N)} = R_{m(n(N))}/\mathfrak{m}_{R_{m(n(N))}}^N$ and $D_{n(N)} = R_{n(N),m(n(N))}/\mathfrak{m}_{R_{n(N),m(n(N))}}^N$. Going modulo $N^{\text{th}}$-powers of maximal ideals in the diagram above we get another commutative diagram with exact rows:

$$\begin{array}{ccccccc} C_{n(N)}^r & \longrightarrow & C_{n(N)} & \longrightarrow & R/\mathfrak{m}_R^N & \longrightarrow & 0 \\ \downarrow & & \downarrow & & \downarrow & & \\ D_{n(N)}^r & \longrightarrow & D_{n(N)} & \longrightarrow & \widetilde{R}_{n(N),m(n(N))}/\mathfrak{m}_{\widetilde{R}_{n(N),m(n(N))}}^N & \longrightarrow & 0. \end{array}$$

By our choice of $n(N)$ the first two vertical maps are isomorphisms, so that

$$R/\mathfrak{m}_R^N \xrightarrow{\sim} \widetilde{R}_{n(N),m(n(N))}/\mathfrak{m}_{\widetilde{R}_{n(N),m(n(N))}}^N,$$

for all $N > 0$. Taking limits, we see that $R \xrightarrow{\sim} \widetilde{R}_\infty$. This completes the proof of the isomorphism criterion. Now

$$M = \varprojlim M/p^n M = \varprojlim_n M_{m(n)}/(I_n + \mathfrak{a}_{Q_{m(n)}})M_{m(n)} = L/(S_1, \dots S_r)L.$$

But $L$ is a free $R_\infty$-module and $\widetilde{R}_\infty = R_\infty/(S_1, \dots S_r)R_\infty = T$. We conclude that $M$ is a free $T$-module. This completes the proof of the freeness criterion as well. □

## References

[1] F. Diamond. The Taylor–Wiles construction and multiplicity one. *Invent. Math.* **128** no. 2 (1997) 379–391
[2] K. Fujiwara. Deformation rings and Hecke-algebras in the totally real case. *Preprint* (1996)
[3] E. Ghate. On the freeness of the integral cohomology groups of Hilbert-Blumenthal varieties as Hecke-modules. *In preparation* (2003)
[4] H. Hida. *Modular Forms and Galois Cohomology*. Cambridge University Press, Cambridge (2000)
[5] H. Matsumura. *Commutative Ring Theory*. Cambridge University Press (1986)
[6] C. Skinner and A. Wiles. Residually reducible representations and modular forms. *Inst. Hautes Études Sci. Publ. Math.* **89** (2000) 5–126
[7] R. Taylor and A. Wiles. Ring-theoretic properties of certain Hecke algebras. *Ann. of Math.* **141** no. 3 (1995) 553–572
[8] A. Wiles. Modular elliptic curves and Fermat's last theorem. *Ann. of Math.* **141** no. 3 (1995) 443–551

Proc. Int. Conf. – Number Theory No. 1, 2004, pp. 15–29.

# The Mellin Transform of the Fourth Power of Riemann's Zeta-Function

Matti Jutila

University of Turku, Department of Mathematics, FIN-20014 Turku, Finland
e-mail: jutila@utu.fi

*Dedicated to Professor M. V. Subba Rao on his 80th birthday*

## 1. Introduction

The moments

$$I_k(T) = \int_0^T \left|\zeta\left(\frac{1}{2}+it\right)\right|^{2k} dt \ \ (k = 1, 2 \dots) \tag{1.1}$$

of Riemann's zeta-function, playing an important role in arithmetical applications, are also of interest for our understanding of the zeta-function itself. Various approaches to these moments have been devised in the past such as the classical argument based on approximate functional equations. Other natural tools for the same purpose are integral transforms, in particular the Laplace and Mellin transforms. We developed a "Laplace-method" in a series of papers ([J2], [J4], [J5]) for $k = 1$ and 2. On the other hand, a "Mellin-method", closely analogous to the standard Perron integral method in the theory of arithmetic functions, was introduced by Y. Motohashi (see [M2] and [M3], Sec. 5.3), and it is the subject of the present paper.

The function

$$\mathcal{Z}_k(s) = \int_1^\infty \left|\zeta\left(\frac{1}{2}+ix\right)\right|^{2k} x^{-s}\, dx \tag{1.2}$$

is a modified Mellin transform of the $2k$'th zeta-power. Actually $\mathcal{Z}_k(1-s)$ would represent the Mellin transform in the usual sense, but we prefer the definition (1.2) suggesting the analogy of $\mathcal{Z}_k(s)$ with the generating Dirichlet series of an arithmetical function. Motohashi showed the meromorphy of $\mathcal{Z}_2(s)$ in the whole complex plane and determined its poles in the half-plane $\sigma > 0$ with an application to omega-theorems for the error term $E_2(T)$ in his mind; as usual, $E_2(T)$ denotes the error term in the asymptotic formula for $I_2(T)$. In [M2], the important result $E_2(T) = \Omega_\pm(\sqrt{T})$ was proved, and since the function $\mathcal{Z}_2(s)$ determines $\left|\zeta\left(\frac{1}{2}+it\right)\right|^4$ by the Mellin inversion, one may expect in principle other applications as well.

Besides the qualitative problem concerning the analytic nature of $\mathcal{Z}_2(s)$, also the quantitative question about its order of magnitude turned out to be of relevance in the study of A. Ivić [I2] of more delicate oscillatory properties of the error term $E_2(T)$. This growth problem was tackled in [IJM] with the main result that $\mathcal{Z}_2(s)$ is *of polynomial order* on vertical lines in the half-plane $\sigma > -\frac{1}{2}$ if the poles are of course excluded with small neighborhoods. In the same paper, the function $\mathcal{Z}_1(s)$ was also discussed, by use of Atkinson's classical formula for $I_1(T)$. Another approach would be to modify Motohashi's argument for $\mathcal{Z}_2(s)$. In [J6], we presented recently a *third* approach to $\mathcal{Z}_1(s)$ - and in fact more generally to $\mathcal{Z}_k(s)$ - on interpreting it first as a complex integral, and manipulating it then suitably to end up with something similar to the function $\zeta^2(s)$. The affinity between $\mathcal{Z}_1(s)$ and $\zeta^2(s)$ was less explicitly visible already in [IJM].

Thus, the underlying idea in [J6] was to connect the function $\mathcal{Z}_1(s)$ with $\zeta^2(s)$ whose meromorphic continuation is classically known, and we pointed out that analogously the function $\mathcal{Z}_2(s)$ can be related to the additive divisor problem zeta-function, which appears in somewhat different versions in the literature, such as

$$\zeta_f(s) = \sum_{n=1}^{\infty} d(n)d(n+f)\left(n+\frac{1}{2}f\right)^{-s} \tag{1.3}$$

or

$$\tilde{\zeta}_f(s) = \sum_{n=1}^{\infty} d(n)d(n+f)(n+f)^{-s}. \tag{1.4}$$

In addition, their spectral theoretic variant $\zeta_f^*(s)$, to play a key role in this paper, will be discussed and applied in Sec. 7. By its very definition, the last mentioned function is meromorphic in the whole plane, and this fact implies the meromorphic continuation of the other two variants as well (see [S], [TV], [J1], [J3]) by certain comparison arguments. Another consequence of the spectral decomposition of $\zeta_f^*(s)$ is the polynomial growth of all these three functions on vertical lines.

Though our method in [J6] could be applied to $\mathcal{Z}_2(s)$ - and in fact we worked out this argument in an earlier version of this paper - it turned out that it is simpler to start from the Laplace transform and to utilize the connection between the Laplace and Mellin transforms. True, this argument without further elaboration gives the meromorphic continuation only in the half-plane $\sigma > 0$, whereas the alternative argument would work in the whole plane. An arithmetical formula for the Laplace transform for $\left|\zeta\left(\frac{1}{2}+it\right)\right|^4$ was given by F. V. Atkinson [A], and in [J4] we generalized his formula to cusp form $L$-functions. In this result (see Lemma 1 below) there is a correction term $L_0(p)$ involved for which we would need a suitable explicit expression to extend the meromorphy of $\mathcal{Z}_2(s)$ to the left half-plane. However, the latter case is mainly of theoretical rather than practical interest. Therefore, and to avoid extra complications, we are going to consider $\mathcal{Z}_2(s)$ only in the right half-plane.

If we follow [J6], we have to apply Voronoi's summation formula to get a sum of the additive divisor problem type, whereas the Voronoi summation is already hidden, in the equivalent form of the functional equation for $\zeta^2(s)$, in Atkinson's proof of his formula for the Laplace transform. That is why the present argument is simpler.

We now summarize the main properties of the function $\mathcal{Z}_2(s)$. The (standard) spectral theoretic notations $\kappa_j$ and $H_j(s)$ will be explained in Sec. 7.

**Theorem.** *The function $\mathcal{Z}_2(s)$ is meromorphic in the whole complex plane, and it is of polynomial order in $|s|$ on vertical lines if the poles are excluded by small neighborhoods. In particular, for any fixed number $c \in (1/2, 1)$ we have*

$$\mathcal{Z}_2(c+it) \ll (|t|+1)^{1/2}. \tag{1.5}$$

*The poles of $\mathcal{Z}_2(s)$ in the half-plane $\sigma > 0$ are: a pole of order five at $s = 1$, simple poles $\frac{1}{2} + i\kappa_j$ with the residue at $\frac{1}{2} + i\kappa$ being*

$$(\pi/2)^{1/2}\left(2^{-i\kappa}\frac{\Gamma\left(\frac{1}{2}\left(\frac{1}{2}-i\kappa\right)\right)}{\Gamma\left(\frac{1}{2}\left(\frac{1}{2}+i\kappa\right)\right)}\right)^3 \Gamma(2i\kappa)\cosh(\pi\kappa)\sum_{\kappa_j=\kappa}\alpha_j H_j^3\left(\frac{1}{2}\right), \tag{1.6}$$

*and finally the points $\rho/2$ with $\rho$ running over the complex zeros of Riemann's zeta-function.*

The residue (1.6) was calculated by Motohashi (see [M2], or [M3], Theorem 5.4). From the estimate (1.5), due to Ivić [I3], one may deduce (see Sec. 2) in yet another way the bound

$$E_2(T) \ll T^{2/3+\varepsilon}. \tag{1.7}$$

The main point of the present paper is the novelty of the method which is "functional" in the sense that it applies as such to cusp form $L$-functions as well.

More specifically, we are going to prove certain parts of the theorem, namely the assertions concerning the meromorphy and the poles of $\mathcal{Z}_2(s)$ in the half-plane $\sigma > 0$, and the estimate (1.5) also follows up to terms of the form

$$\int_1^Y \left|\zeta\left(\frac{1}{2}+it\right)\right|^4 x^{-s}\,dx \tag{1.8}$$

with $Y \ll (|t|+1)^{2+\varepsilon}$. Though this function could be shown to be of the desired order, we do not go into details because such terms in fact turn out to be irrelevant for a proof of (1.7) which was the main motivation and test for (1.5). Partially "new" (though implicit) results will be the meromorphy and polynomial growth of the Mellin transform of the square of a Hecke $L$-function attached to a cusp form for the full modular group. The present method yields this only in the right half-plane, but following [J6] one could remove this restriction. In particular, analogues of the estimate (1.5) imply analogues of (1.7) for cusp form $L$-functions. We gave a unified proof of the latter result already in [J4], but the present method appears to be simpler. It should be noted that the meromorphic continuation of the Mellin transform in the case of holomorphic cusp forms was shown by Motohashi [M1], but the case of non-holomorphic forms has been an open problem.

## 2. Connection between $\mathcal{Z}_2(s)$ and $E_2(X)$

Since the parameter $T$ will have a special and fixed meaning in this paper, we prefer to introduce another parameter $X$ and discuss the error term $E_2(X)$; in this context, $T$ will lie in the range (2.2) below.

An obvious device for expressing the moment $I_2(X)$ (and hence the error term $E_2(X)$) in terms of the function $\mathcal{Z}_2(s)$ would be an analogue of Perron's formula, but to follow the standard argument one would need a good bound for $\left|\zeta\left(\frac{1}{2}+ix\right)\right|^4$. This difficulty can be avoided by a certain averaging argument as shown by A. Ivić [I3]. Let $1 \ll H \leq \frac{1}{4}X$ and let $f(x)$ be a non-negative smooth function supported in $[0, H]$ such that $f(x) = 1$ for $\frac{1}{4}H \leq x \leq \frac{3}{4}H$. Then, for some positive constants $C_1$ and $C_2$, and with $\frac{1}{2} < c < 1$, we have (see (4.6) in [I3])

$$E_2(X) \leq \frac{C_1}{2\pi i H} \int_{(c)} \frac{\mathcal{Z}_2(s)}{s} \int_0^H f(x)(X+x)^s \, dx \, ds + C_2 H \log^4 X, \tag{2.1}$$

and an analogous lower bound holds as well. The $s$-integral can be truncated to the finite segment $|t| \leq X^{1+\varepsilon}H^{-1}$ with a negligible error. Let now $T < X^{1+\varepsilon}H^{-1}$ and consider the $s$-integral in (2.1) over $T \leq t \leq 2T$. The relevant values of $H$ exceed $X^{1/2}$ since the best one can hope is the estimate $E_2(X) \ll X^{1/2+\varepsilon}$, and therefore we suppose that

$$T^{2+\eta} \leq X \leq T^A \tag{2.2}$$

for some positive constants $\eta$ and $A$ with $\eta$ small.

We note now that the "partial integral" (1.8) of $\mathcal{Z}_2(s)$, with $Y \ll T^{2+\varepsilon}$, can be ignored in a range where $t \asymp T$, as far as the estimation of $E_2(X)$ is concerned. To see this, move in (2.1) the integration over $[T, 2T]$ sufficiently far to the left and estimate the integral (1.8) trivially by absolute values.

Suppose now that

$$\int_0^T |\mathcal{Z}_2(c+it)|^2 \, dt \ll T^a \tag{2.3}$$

for any fixed number $c \in \left(\frac{1}{2}, 1\right)$, the implied constant depending on $c$. Then, with $H = X^{a/(1+a)}$, we deduce from (2.1) that

$$E_2(X) \ll X^{a/(1+a)+\varepsilon}. \tag{2.4}$$

By (1.5), the value $a = 2$ is admissible, which gives the above mentioned bound $E_2(X) \ll X^{2/3+\varepsilon}$, whereas the conjectural value $a = 1 + \varepsilon$ (Conjecture 1 in [I3]) would give the essentially best possible bound $E_2(X) \ll X^{1/2+\varepsilon}$.

## 3. The Laplace and Mellin transforms

Our starting point is a familiar relation between the Laplace transform

$$L(p) = \int_1^\infty \left|\zeta\left(\frac{1}{2}+ix\right)\right|^4 e^{-px} \, dx \tag{3.1}$$

and the Mellin transform $\mathcal{Z}_2(s)$: obviously we have

$$\mathcal{Z}_2(s) = \frac{1}{\Gamma(s)} \int_0^{\infty} L(p) p^{s-1}\, dp. \tag{3.2}$$

To make use of this formula, we invoke the following formula for $L(p)$ (see [J4], Lemma 1).

**Lemma 1.** *Let $\theta \in (0, \pi)$ be fixed. Then, for $0 < Re\ p \leq \theta$ and $|Im\ p| \ll 1$, we have*

$$L(p) = 2i \sum_{m,n=1}^{\infty} d(m)d(n)L_{m,n}(p) + L_0(p), \tag{3.3}$$

*where*

$$L_{m,n}(p) = \frac{e(-me^{-ip/2} + ne^{ip/2})}{-me^{-ip/2} + ne^{ip/2}} \tag{3.4}$$

*and*

$$L_0(p) \ll \log^3 \left(1 + (Re\ p)^{-1}\right). \tag{3.5}$$

In [J4], the lower limit of integration in the definition of $L(p)$ was actually 0 instead of 1, but this affects only the correction function $L_0(p)$ in (3.3).

By the well-known estimate for the fourth moment of the zeta-function, the function $\mathcal{Z}_2(s) = \mathcal{Z}_2(\sigma + it)$ is holomorphic in the half-plane $\sigma > 1$. For the desired meromorphic continuation, we introduce a parameter $T \geq 2$ and intend to extend $\mathcal{Z}_2(s)$ meromorphically to the domain $D(T)$ defined by the conditions

$$\sigma > 0 \text{ and } t \in \begin{cases} [0, 2) & \text{for} \quad T = 2, \\ (T/2, T) & \text{for} \quad T > 2. \end{cases} \tag{3.6}$$

Returning to (3.2), we note that the factor $1/\Gamma(s)$ grows exponentially as a function of $t$, and for this sake we turn the path of integration to the ray $\arg p = \alpha$ with

$$\alpha = \pi/2 - 1/T. \tag{3.7}$$

Then $p = ue^{i\alpha}$ for real $u \geq 0$, and the factor $p^{s-1}$ now compensates the exponential growth of the inverse gamma-factor for $s \in D(T)$. We express $L(p)$ by Lemma 1 for $0 < u \leq T^{-1-\delta}$, where $\delta$ is a small positive constant, and then integrals of the form

$$I_{m,n}(s) = \frac{2i}{\Gamma(s)} \int_0^{p_0} L_{m,n}(p) p^{s-1}\, dp \tag{3.8}$$

appear with $p_0 = T^{-1-\delta} e^{i\alpha}$. On the other hand, for $u > T^{-1-\delta}$, we use simply the definition (3.1) for $L(p)$. Separating the diagonal and nondiagonal terms in (3.3), we get

$$\mathcal{Z}_2(s) = \sum_{j=0}^{3} \mathcal{Z}_{2j}(s), \tag{3.9}$$

where

$$\mathcal{Z}_{20}(s) = \frac{1}{\Gamma(s)} \int_0^{p_0} L_0(p) p^{s-1}\, dp,$$

$$\mathcal{Z}_{21}(s) = \frac{1}{\Gamma(s)} \int_{p_0}^{\infty} \left( \int_1^{\infty} \left| \zeta\left(\frac{1}{2} + ix\right) \right|^4 e^{-px}\, dx \right) p^{s-1}\, dp, \tag{3.10}$$

$$\mathcal{Z}_{22}(s) = \sum_{n=1}^{\infty} d^2(n) I_{n,n}(s), \tag{3.11}$$

$$\mathcal{Z}_{23}(s) = \sum_{m \neq n} d(m) d(n) I_{m,n}(s). \tag{3.12}$$

First of all, the term $\mathcal{Z}_{20}(s)$ is easy, for by (3.5) it is holomorphic in $D(T)$, and estimating the gamma-function by Stirling's formula we see that in the same domain it holds

$$\mathcal{Z}_{20}(s) \ll T^{1/2-\sigma+\varepsilon} \text{ for } \sigma \geq \sigma_0 > 0. \tag{3.13}$$

Next, by (3.10), the function $\mathcal{Z}_{21}(s)$ can be written in terms of the incomplete gamma-function as follows:

$$\mathcal{Z}_{21}(s) = \frac{1}{\Gamma(s)} \int_1^{\infty} \left| \zeta\left(\frac{1}{2} + ix\right) \right|^4 x^{-s} \Gamma(s, p_0 x)\, dx.$$

Since the ratio $\Gamma(s, p_0 x)/\Gamma(s)$, bounded for all $x$, is rapidly decreasing for $x \gg T^{2+\delta}$, the integral over this range can be estimated directly. Then we have a similar finite part of $\mathcal{Z}_2(s)$ to deal with as in (1.8), and its contribution to $E_2(X)$ can be estimated as explained in the preceding section.

At this stage, we make a preliminary remark concerning the analysis of the remaining ingredients of $\mathcal{Z}_2(s)$. Supposing that we already know the analytic continuation of $\mathcal{Z}_2(s)$ to that part of $D(T)$ lying to the right of the half-line, we want to show the estimate (1.5), up to some terms of the form (1.8). Since the polynomial growth of $\mathcal{Z}_2(s)$ will be clear in principle from our argument, we may suppose that $T$ is sufficiently large in this context. Let $s_0 = c + it_0$ with $c > 1/2$ and $T/2 \leq t_0 \leq T$, and write

$$\mathcal{Z}_2(s_0) = \frac{1}{2\pi i} \int_{\gamma} \frac{\mathcal{Z}_2(s) U^{s-s_0}}{s - s_0} e^{(s-s_0)^2}\, ds, \tag{3.14}$$

where $U = T^{2+2\delta}$ and $\gamma$ is a rectangular contour with the vertices $b + i(t_0 \pm \log T)$, $d + i(t_0 \pm \log T)$ for some constants $b \in \left(\frac{1}{2}, c\right)$ and $d$ slightly exceeding 1.

In the integral over the right side of $\gamma$, the original definition (1.2) of $\mathcal{Z}_2(s)$ is valid, and we split up the $x$-integral into two parts over the ranges $1 \leq x \leq T^{2+3\delta}$ and $x > T^{2+3\delta}$. The latter part gives a negligible contribution to the integral (3.14), as can be seen on moving the integration sufficiently far to the right. The contribution of the former part is treated on moving the integration to a line a bit to the right of the line $\sigma = c$, and then we end up essentially with "partial integrals" of the type (1.8). The integrals over the horizontal sides of $\gamma$ are small, taken that we know that $\mathcal{Z}_2(s)$ is of polynomial order. Thus, all in all, it suffices to consider integrals of the type

$$\frac{1}{2\pi i}\int_{b+i(t_0-\log T)}^{b+i(t_0+\log T)} \frac{f(z)U^{s-s_0}}{s-s_0} e^{(s-s_0)^2}\, ds \tag{3.15}$$

for various terms $f(z)$ to occur in the decomposition of $\mathcal{Z}_2(s)$. An advantage of this formula is that the factor $U^{s-s_0}$ decays rapidly if the integration is moved to the right.

## 4. The integrals $I_{m,n}(s)$

To simplify the analysis of the remaining functions $\mathcal{Z}_{22}(s)$ and $\mathcal{Z}_{23}(s)$ in (3.9), we transform the integrals $I_{m,n}(s)$ substituting a new variable $q$ by the relation

$$ne^{ip/2} - me^{-ip/2} = f + (m+n)iq/2,$$

where $f = n - m$. Thus $p \approx q$ in the first approximation if $p$ is small, and solving this equation for $p$ we may expand it into a power series in $q$:

$$p = q + rq^2 + \cdots ,$$

where $r = if/(4(m+n))$, and in general the coefficients are polynomials in $f/(m+n)$.
Hence

$$I_{m,n}(s) = \frac{2i}{\Gamma(s)}\int_0^{q_0} \frac{e^{-\pi(m+n)q}q^{s-1}}{f+(m+n)iq/2}(1+rq+\cdots)^{s-1}(1+2rq+\cdots)\,dq, \tag{4.1}$$

where $q_0$ corresponds to $p_0$ and the integral is taken over the line segment $[0, q_0]$. Note that $\pi/2 - \arg q_0 \asymp 1/T$ similarly to (3.7), so that the slope of the path of integration is not essentially changed.

To simplify the integrand, note that the power $(\cdots)^{s-1}$ can be expanded into a rapidly convergent power series since $|q| \le T^{-1-\delta}$. Thus the product of the two factors involving power series can be written as a single power series with leading term 1, and the other terms yield functions in which $s$ is shifted by a natural number. It suffices to deal with the leading term, for the others are analogous but easier. The series can be truncated after finitely many terms, for the reminder gives a function which is holomorphic and small in $D(T)$.

Next we extend the $q$-integral to infinity and correct the error. The terms with $m + n \ge T^{2+2\delta}$ make only a small contribution, and the remaining finitely many pairs $m, n$ give a holomorphic function in $D(T)$. Furthermore, the contribution of the last mentioned correction terms to the integral (3.15) is negligible, as can be seen on moving the integration to the left and taking into account that then $(qU)^s/\Gamma(s)$ decays exponentially as a function of decreasing $\sigma$ if $|q| \ge |q_0|$. Thus, all in all, we have reduced the integral (4.1) with a manageable error to

$$J_{m,n}(s) = \frac{2i}{\Gamma(s)}\int_0^{\infty} \frac{e^{-\pi(m+n)q}q^{s-1}}{f+(m+n)iq/2}\,dq. \tag{4.2}$$

The integral $J_{m,n}(s)$ can be expressed in terms of the incomplete gamma-function $\Gamma(s,x)$, and also in terms of generalized Fresnel cosine integrals

$$C(x,s)=\int_x^\infty y^{s-1}\cos y\,dy. \tag{4.3}$$

Such formulae, stated in the next lemma, also give analytic continuations of our integrals.

**Lemma 2.** *Let $m, n$ be natural numbers and put $f=n-m$, $f_0=|f|$. Then*

$$J_{n,n}(s)=2(2\pi)^{1-s}n^{-s}/(s-1), \tag{4.4}$$

*and for $m\neq n$ we have*

$$\begin{aligned}J_{m,n}(s)+J_{n,m}(s)&=2^{1+s}if_0^{s-1}(m+n)^{-s}\\&\quad\times\left(e^{-(\pi/2)is}\Gamma(1-s,-2\pi if_0)-e^{(\pi/2)is}\Gamma(1-s,2\pi if_0)\right).\end{aligned} \tag{4.5}$$

*Moreover, in the latter case,*

$$J_{m,n}(s)+J_{n,m}(s)=2^{2+s}f_0^{s-1}C(2\pi f_0,1-s)(m+n)^{-s}. \tag{4.6}$$

*Proof.* The equation (4.4) is immediate. To the non-diagonal integrals, we apply the formula (see [EMOT], §9.3, eq. (3))

$$\int_0^\infty\frac{x^{\nu-1}e^{-\mu x}}{x+\beta}\,dx=\Gamma(\nu)\beta^{\nu-1}e^{\beta\mu}\Gamma(1-\nu,\beta\mu),$$

valid for $|\arg\beta|<\pi$, $\mathrm{Re}\mu>0$, and $\mathrm{Re}\,\nu>0$. This gives

$$\begin{aligned}J_{m,n}(s)&=4(-2if)^{s-1}\Gamma(1-s,-2\pi if)(m+n)^{-s}\\&=\mathrm{sgn}(f)2^{1+s}if_0^{s-1}e^{-(\pi/2)\mathrm{sgn}(f)is}\Gamma(1-s,-2\pi if)(m+n)^{-s},\end{aligned}$$

which proves (4.5). Finally, for a proof of (4.6), we apply the relation (see [EMOT], §9.10, eq. (10))

$$C(x,s)=\frac{1}{2}e^{-\pi is/2}\Gamma(s,ix)+\frac{1}{2}e^{\pi is/2}\Gamma(s,-ix).$$

For subsequent reference, we further quote from [EMOT] some basic properties of the function $C(x,s)$, writing $(a)_\nu=a(a-1)\cdots(a-\nu+1)$ as usual.

**Lemma 3.** *We have*

$$C(0,s)=\cos\left(\frac{1}{2}\pi s\right)\Gamma(s). \tag{4.7}$$

*Also, for large $x$ exceeding $|s|$, the following asymptotic formula holds:*

$$C(x,s)=-x^s(P(x)\sin x+Q(x)\cos x),$$

*where*

$$P(x) = \sum_{m=0}^{M-1} \frac{(-1)^m (1-s)_{2m}}{x^{2m+1}} + O(|x|^{-2M-1})$$

*and*

$$Q(x) = \sum_{m=1}^{M} \frac{(-1)^m (1-s)_{2m-1}}{x^{2m}} + O(|x|^{-2M-2}).$$

## 5. The function $\mathcal{Z}_{22}(s)$

In the preceding section, we found that instead of the function $\mathcal{Z}_{22}(s)$ in its original form (3.11), it suffices to consider the modified function

$$\sum_{n=1}^{\infty} d^2(n) J_{n,n}(s).$$

By (4.4), this equals

$$2(2\pi)^{1-s}(s-1)^{-1} \sum_{n=1}^{\infty} d^2(n) n^{-s} = 2(2\pi)^{1-s}(s-1)^{-1} \frac{\zeta^4(s)}{\zeta(2s)}.$$

The last mentioned function has a pole of the order five at $s = 1$, and it has also poles at the points $\rho/2$, where $\rho$ runs over the complex zeros of Riemann's zeta-function. Moreover, this function is bounded if $\sigma$ is bounded below by a constant exceeding $\frac{1}{2}$ and a neighborhood of the point 1 is excluded.

## 6. The function $\mathcal{Z}_{23}(s)$: reduction to $\tilde{\zeta}_f(s)$

Again it suffices to study a simplified version, that is

$$\sum_{m \neq n} d(m) d(n) J_{m,n}(s),$$

of our function. By (4.6) and (1.3), this can be written as

$$4 \sum_{f=1}^{\infty} f^{s-1} C(2\pi f, 1-s) \zeta_f(s). \tag{6.1}$$

The behavior of $C(2\pi f, 1-s)$ depends on the size of $f$ relative to $T$, for the corresponding integral (4.3) has a saddle point at $y = t$ which lies, in our case, in the range of integration only if $f < t/2\pi$. Then $C(2\pi f, 1-s)$ is comparable with

$$C(0, 1-s) = \sin\left(\frac{1}{2}\pi s\right) \Gamma(1-s),$$

which follows from (4.7). In any case, by the saddle point method and elementary estimations, we have

$$C(2\pi f, 1-s) \ll T^{1/2-\sigma} + f^{1-\sigma}T^{-1}$$

for all $f \geq 1$ and all $s \in D(T)$. On the other hand, if $f$ exceeds $T$, then it is preferable to use the asymptotic formula (4.8) implying that

$$C(2\pi f, 1-s) = s(2\pi f)^{-1-s} + O(f^{-3-\sigma}) \tag{6.2}$$

for $f \geq CT$, where $C$ is a sufficiently large constant and the $O$-constant depends on $T$. The estimate $C(2\pi f, 1-s) \ll Tf^{-1-\sigma}$ given by the leading term holds in fact uniformly for $T$, as one can verify by partial integration and the first derivative test.

To normalize the coefficients $C(2\pi f, 1-s)$, define

$$A(f,s) = C(2\pi f, 1-s)\left(\Gamma(1-s)\sin\left(\frac{1}{2}\pi s\right)\right)^{-1},$$

so that $A(f,s) \approx 1$ is a certain initial range for $f$ and always $A(f,s) \ll 1$, whereas by (6.2) and the subsequent remark we have

$$A(f,s) \ll T^{1/2+\sigma} f^{-1-\sigma} \text{ for } f \geq CT. \tag{6.3}$$

The (6.1) takes the form

$$4\Gamma(1-s)\sin\left(\frac{1}{2}\pi s\right)\sum_{f=1}^{\infty} A(f,s) f^{s-1}\zeta_f(s). \tag{6.4}$$

As a minor modification, to be motivated in the next section, we want to replace $\zeta_f(s)$ here by the function $\tilde{\zeta}_f(s)$ defined in (1.4); this means replacing (6.4) by the function

$$\tilde{\mathcal{Z}}_{23}(s) = 4\Gamma(1-s)\sin\left(\frac{1}{2}\pi s\right)\sum_{f=1}^{\infty} A(f,s) f^{s-1}\tilde{\zeta}_f(s). \tag{6.5}$$

To compare the functions (6.4) and (6.5), let us consider their difference

$$4\Gamma(1-s)\sin\left(\frac{1}{2}\pi s\right)\sum_{f=1}^{\infty} A(f,s) f^{s-1} \times \sum_{n=1}^{\infty} d(n)d(n+f)\left(\left(n+\frac{1}{2}f\right)^{-s} - (n+f)^{-s}\right). \tag{6.6}$$

By elementary estimations and the above mentioned bounds for $A(f,s)$, this is clearly holomorphic in $D(T)$, where the functions (6.4) and (6.5) are thus of the same analytic nature.

For a closer study of (6.6), note that if $f/(n+f) \leq T^{-1-\delta}$, then the difference in the brackets can be expanded into a rapidly convergent series involving $(n+f)^{-s-j} f^j$ for $j = 1, 2, \ldots$. Thus, up to the missing terms with $n + f < fT^{1+\delta}$, we get shifted functions of the type (6.5), and the expansion can be truncated to a sum over finitely many values of $j$. The missing terms can be inserted back to get complete series $\tilde{\zeta}_f(s+j)$, for their contribution to the integral (3.15) is anyway small, as can be seen on moving the integration sufficiently far to the left. By these remarks, we may now take the function (6.5) as the basis for a further analysis and the analytic continuation of the function $\mathcal{Z}_{23}(s)$.

## 7. Meromorphic continuation of $\mathcal{Z}_2(s)$

We have seen so far that to continue the function $\mathcal{Z}_2(s)$ meromorphically to the right half-plane, it remains to do the same for the function $\tilde{\mathcal{Z}}_{23}(s)$ defined in (6.5). In fact, this function can be continued to the whole plane.

The key for the meromorphic continuation is the observation that there is a meromorphic function $\zeta_f^*(s)$, closely related to $\tilde{\zeta}_f(s)$, which has both an arithmetic and a spectral theoretic representation (see [TV], [J1], [J3]). Accordingly, as another modification of the function $\mathcal{Z}_{23}(s)$, we introduce the function

$$\mathcal{Z}_{23}^*(s) = 4\Gamma(1-s)\sin\left(\frac{1}{2}\pi s\right)\sum_{f=1}^{\infty} A(f,s)f^{s-1}\zeta_f^*(s), \tag{7.1}$$

which will be seen to be an approximation to $\tilde{\mathcal{Z}}_{23}(s)$. At this stage, we make a digression to recall the definition of $\zeta_f^*(s)$ and to summarize its salient properties.

We use the standard notation of the spectral theory of automorphic functions (see [M3]). The eigenvalues in the discrete spectrum of the hyperbolic Laplacian are written as $1/4+\kappa_j^2$ in ascending order, $\rho_j(n)$ stands for the Fourier coefficients of the $j$th Maass wave form and $\alpha_j = |\rho_j(1)|^2/\cosh(\pi\kappa_j)$; the corresponding Hecke-eigenvalues are $t_j(n)$ with $\rho_j(n) = t_j(n)\rho_j(1)$ for $n \geq 1$, and the respective Hecke $L$-function is

$$H_j(s) = \sum_{n=1}^{\infty} t_j(n)n^{-s}.$$

It is well-known that there are $\ll K$ numbers $\kappa_j$ in any interval $[K, K+1]$ with $K \geq 1$. Moreover, the first of these satisfies $\kappa_1 > 3.815$ (see [M3], Lemma 1.4).

Putting now $z_j = 1/2 + i\kappa_j$ and $\xi(s) = \pi^{-s/2}\Gamma(s/2)\zeta(s)$, define

$$\begin{aligned} \zeta_f^*(s) &= Z_f(s) \qquad\qquad (7.2)\\ &+ \frac{f^{1/2-s}}{4^s\Gamma^4(s/2)}\sum_{j=1}^{\infty}\alpha_j\cosh(\pi\kappa_j)t_j(f)H_j^2\left(\frac{1}{2}\right)|\Gamma(z_j/2)|^4\Gamma(s-z_j)\Gamma(s-\overline{z_j})\\ &+ \frac{\pi f^{1/2-s}}{4^s\Gamma^4(s/2)}\int_{-\infty}^{\infty}\frac{\sigma_{2iu}(f)\xi^4\left(\frac{1}{2}+iu\right)\Gamma\left(s-\frac{1}{2}+iu\right)\Gamma\left(s-\frac{1}{2}-iu\right)}{f^{iu}|\Gamma\left(\frac{1}{2}+iu\right)|^2|\zeta(1+2iu)|^2}\,du, \end{aligned}$$

where $Z_f(s)$ is a meromorphic function with a triple pole at $s=1$. The last mentioned function can be written in terms of $f$ and the gamma-function (see [TV], Theorem 1), and we have

$$Z_f(s) \ll f^{1-\sigma}(f(|t|+1))^{\varepsilon} \tag{7.3}$$

if $\sigma$ is bounded and small neighborhoods of the poles $1,\ 0,\ -1,\ldots$ of $Z_f(s)$ are excluded.

The meromorphy of the second ingredient in (7.2) is also clear by familiar properties of the gamma-function and the following spectral mean value results (see [K], Theorem 6, and [M3], Theorem 3.4):

$$\sum_{\kappa_j \le K} \alpha_j t_j^2(f) = \frac{1}{\pi^2} K^2 + O(K \log K + K f^{\varepsilon} + f^{1/2+\varepsilon}), \tag{7.4}$$

$$\sum_{\kappa_j \le K} \alpha_j H_j^4\left(\frac{1}{2}\right) \ll K^2 \log^{15} K. \tag{7.5}$$

The meromorphic continuation of the last term in (7.2) can be effected by writing it as a complex integral with respect to the variable $w = 1/2 + iu$ as follows:

$$\frac{\pi f^{1/2-s}}{4^s \Gamma^4(s/2) i} \int_{(1/2)} \frac{\sigma_{2w-1}(f)\xi^4(w)\Gamma(s+w-1)\Gamma(s-w)}{f^{w-1/2}\Gamma(w)\Gamma(1-w)\zeta(2w)\zeta(2-2w)}\, dw.$$

Suppose first that $1/2 < \sigma < 1$ and move the integration to the line $\operatorname{Re} w = 1$ with a small indentation to the left at the point $w = 1$. We encounter poles at the points $w = s$ and $w = 1 - \rho/2$ with $\rho$ running over the complex zeros of Riemann's zeta-function. The new integral defines a holomorphic function in the whole strip $0 < \sigma < 1$, and the residues at the poles are meromorphic functions of $s$ in this strip. In this way, the meromorphic continuation of the last term in (7.2) to this strip, and in fact to the right half-plane, is established. Letting now $0 < \sigma < 1/2$, shift the integration back to the half-line observing that the residues at the poles $1 - \rho/2$ are cancelled. On the other hand, the previous residual term

$$\frac{\pi^{3/2-2s}\sigma_{1-2s}(f)\zeta^4(s)\Gamma(s-1/2)}{2\zeta(2s)\zeta(2-2s)\Gamma(1-s)}$$

survives and gives poles at the points $\rho/2$. So far we have a meromorphic continuation of $\zeta_f^*(s)$ to the half-plane $\sigma > -1/2$, and the domain of meromorphy can be extended in the same way step by step.

By these remarks and (6.3), the function $\mathcal{Z}_{23}^*(s)$ is meromorphic in $D(T)$ and inherits the poles of $\zeta_f^*(s)$. The meromorphic continuation of $\tilde{\mathcal{Z}}_{23}(s)$ now follows by a comparison argument. To this end, put

$$P_f(s) = \tilde{\zeta}_f(s) - \zeta_f^*(s),$$

and consider the difference

$$\tilde{\mathcal{Z}}_{23}(s) - \mathcal{Z}_{23}^*(s) = 4\Gamma(1-s)\sin\left(\frac{1}{2}\pi s\right)\sum_{f=1}^{\infty} A(f,s) f^{s-1} P_f(s). \tag{7.6}$$

It is known that the function $P_f(s)$, holomorphic in the first place for $\sigma > 1$, admits an analytic continuation to the right half-plane (see [TV], [J1], [J3]). Moreover, we have ([J3], Theorem 2)

$$P_f(s) \ll f^{-\sigma}(f + |t|^{1/2})(f(|t|+1))^{\varepsilon}$$

in any fixed strip $\sigma_0 \le \sigma \le \sigma_1$ with $\sigma_0 > 0$. Hence the function (7.6) is holomorphic and of polynomial order in $D(T)$ for $\sigma \ge \sigma_0 > 0$. Also, we see that the contribution of the function (7.6) to the integral (3.15) (shifted to a line $\sigma = \varepsilon$) is $\ll T^{1/2}$, which is admissible for our purposes.

The function $\mathcal{Z}_{23}^*(s)$, which represents the most significant part of $\mathcal{Z}_2(s)$, decomposes into three parts according to the three terms in (7.2). The contribution of the leading term $Z_f(s)$ is easily shown to be admissible, again by use of the integral (3.15). Also, the contribution of the last term in (7.2) can be estimated satisfactorily by standard results on the zeta-function.

The second term in (7.2), that is the spectral sum, produces the most crucial and interesting part of $\mathcal{Z}_{23}^*(s)$, and thus of $\mathcal{Z}_2(s)$. Let us calculate its residue at the pole $\frac{1}{2} + i\kappa$; of course, the result should be consistent with (1.6).

By (7.1) and (7.2), this residue is equal to

$$2^{1-2i\kappa}\sin\left(\frac{\pi}{4}+\frac{\pi i\kappa}{2}\right)\Gamma\left(\frac{1}{2}-i\kappa\right)\Gamma(2i\kappa)\cosh(\pi\kappa)$$
$$\times\frac{\Gamma^2\left(\frac{1}{2}\left(\frac{1}{2}-i\kappa\right)\right)}{\Gamma^2\left(\frac{1}{2}\left(\frac{1}{2}+i\kappa\right)\right)}\sum_{\kappa_j=\kappa}\alpha_j H_j^2\left(\frac{1}{2}\right)\sum_{f=1}^{\infty}A(f,z_j)t_j(f)f^{-1/2}. \tag{7.7}$$

By familiar formulae for the gamma-function, we have

$$\Gamma\left(\frac{1}{2}-i\kappa\right) = \pi^{-1/2}2^{-1/2-i\kappa}\Gamma\left(\frac{1}{4}-\frac{i\kappa}{2}\right)\Gamma\left(\frac{3}{4}-\frac{i\kappa}{2}\right),$$

$$\Gamma\left(\frac{3}{4}-\frac{i\kappa}{2}\right)\Gamma\left(\frac{1}{4}+\frac{i\kappa}{2}\right) = \frac{\pi}{\sin\left(\frac{\pi}{4}+\frac{\pi i\kappa}{2}\right)}.$$

Hence the residue (7.7) becomes

$$(2\pi)^{1/2}\Gamma(2i\kappa)\cosh(\pi\kappa)\left(2^{-i\kappa}\frac{\Gamma\left(\frac{1}{2}\left(\frac{1}{2}-i\kappa\right)\right)}{\Gamma\left(\frac{1}{2}\left(\frac{1}{2}+i\kappa\right)\right)}\right)^3$$
$$\times\sum_{\kappa_j=\kappa}\alpha_j H_j^2\left(\frac{1}{2}\right)\sum_{f=1}^{\infty}A(f,z_j)t_j(f)f^{-1/2}.$$

If the $f$-sum here is interpreted in the light of the (heuristic) approximation

$$\sum_{f=1}^{\infty}A(f,z_j)t_j(f)f^{-1/2} \approx \frac{1}{2}H_j\left(\frac{1}{2}\right)$$

in the sense of the approximate functional equation, we find an agreement with (1.6).

The contribution of the spectral part of (7.2) to $\mathcal{Z}_2(c+it)$ in (1.5) can be analyzed by (3.5) on moving the integration close to the imaginary axis and taking into account the residues (1.6) at the relevant poles $z_j$. The new integral is $\ll T^{1/2}$ by straightforward estimations. To the sum of the residues, we apply the estimate

$$\sum_{K \le \kappa_j \le K+1} \alpha_j H_j^3\left(\frac{1}{2}\right) \ll K^{1+\varepsilon}$$

due to Ivić [I4] getting again a contribution $\ll T^{1/2}$.

## References


[A] F. V. Atkinson, The mean value of the zeta-function on the critical line, *Proc. London Math. Soc.* (2) **47** (1941) 174–200

[EMOT] A. Erdélyi, W. Magnus, F. Oberhettinger and F. Tricomi, "Higher Trancendental Functions", vol. II, McGraw-Hill, New York, (1953)

[I1] A. Ivić, "Lectures on mean values of the Riemann zeta function", Tata Institute of Fundamental Research, Lectures on Mathematics and Physics, **82** Bombay (1991)

[I2] A. Ivić, On the error term for the fourth moment of the Riemann zeta-function, *J. London Math. Soc.* **60** (2) (1999) 21–32

[I3] A. Ivić, On some conjectures and results for the Riemann zeta-function and Hecke series, *Acta Arith.* **99** (2001) 115–145

[I4] A. Ivić, On sums of Hecke series in short intervals, *J. Théor. Nombres Bordeaux* **13** (2001) 1–16

[IJM] A. Ivić, M. Jutila and Y. Motohashi, The Mellin transform of powers of the Riemann zeta-function, *Acta Arith.* **95** (2000) 305–342

[J1] M. Jutila, The Additive Divisor Problem and Exponential Sums, in "*Advances in Number Theory: Third Conference of the Canadian Number Theory Association*, Kingston 1991", Oxford Univ. Press, (1993) pp. 113–135

[J2] M. Jutila, The fourth moment of Riemann's zeta-function and the additive divisor problem, in "*Analytic Number Theory: Proc. of a Conference in Honor of Heini Halberstam*", **2** Birkhäuser, (1996) 517–536

[J3] M. Jutila, The additive divisor problem and its analogs for Fourier coefficients of cusp forms. I, *Math. Z.* **223** (1996) 435–461

[J4] M. Jutila, Mean values of Dirichlet series via Laplace transform, in "*Analytic Number Theory*" (ed. Y. Motohashi), Cambridge Univ. Press (1997) 169–207

[J5] M. Jutila, Atkinson's formula revisited, in "Voronoi's Impact on Modern Science", Book 1, National Academy of Sciences of Ukraine, Institute of Mathematics, Kyiv (1998) 137–154

[J6] M. Jutila, The Mellin transform of the square of Riemann's zeta-function, *Periodica Math. Hung.* **42** (2001) 179–190

[K] N. V. Kuznetsov, *Petersson's hypothesis for parabolic forms of weight zero and Linnik's hypothesis. Sums of Kloosterman sums* (in Russia), Mat. Sb. **111** (1980) 334–383

[M1] Y. Motohashi, *The mean square of Hecke L-series attached to holomorphic cusp forms*, RIMS Kyoto Univ. Kokyuroku **886** (1994) 214–227

[M2] Y. Motohashi, A relation between the Riemann zeta-fuction and the hyperbolic Laplacian, *Ann. Scuola Norm. Sup. Pisa Cl.Sci* (4) **22** (1995) 299–313

[M3] Y. Motohashi, "Spectral Theory of the Riemann zeta-function", Cambridge Univ. Press (1997)

[S] A. Selberg, On the estimation of Fourier coefficients of modular forms, in "*Proc. Symp. Pure Math.* 8", *American Math. Soc.* (1965) 1–15

[TV] L. A. Takhtadjan and A. I. Vinogradov, *The zeta-function of the additive divisor problem and the spectral decomposition of the automorphic Laplacian* (in Russian), Zap. Nauchn. Sem. LOMI **134** (1984) 84–116

Proc. Int. Conf. – Number Theory No. 1, 2004, pp. 31–37.

# On Dirichlet $L$-Function Values at Rational Arguments

S. Kanemitsu[1], Y. Tanigawa[2] and M. Yoshimoto[3]

[1]Graduate School of Advanced Technology, Kinki University, Iizuka, Fukuoka, 820-8555, Japan
e-mail: kanemitu@fuk.kindai.ac.jp
[2]Graduate School of Mathematics, Nagoya University, Nagoya, 464-8602, Japan
e-mail: tanigawa@math.nagoya-u.ac.jp
[3]Graduate School of Mathematics, Nagoya University, Nagoya, 464-8602, Japan
e-mail: x02001n@math.nagoya-u.ac.jp

**2000 Mathematics Subject Classification:** 11M06, 11M35

*Dedicated to Professor M. V. Subbarao on his eightieth birthday*

## 1. Introduction and statement of results

This is a continuation of our former paper [2] on special values of the multiple Hurwitz zeta-function $\zeta_k(s, \alpha)$ at rational points $s = \frac{2h+1}{2N}$, with $N \geq 1$ and $0 \leq h < N$ being integers. There we proved the results by two different methods. The first uses the Mellin-Vardi decomposition of the multiple Hurwitz zeta-function into a linear combination of ordinary Hurwitz zeta-functions ($k = 1$) as well as the closed form evaluation of the values of the Hurwitz zeta-function at rational points $< 1$ (including those rational points in $(0, 1)$).

The second uses the "quasi modular relation" method by which we mean the Mellin integral method in which the gamma function involved is not the same as one appearing in the functional equation, whereby differentiation is effected by the shift of the variable of integration.

Since Theorem 2.1 [2] indeed gives a closed form for $\zeta\left(-l + \frac{2h+1}{2N}, \alpha\right)$ (or for $\Gamma\left(-l + \frac{2h+1}{2N}\right)\zeta\left(-l + \frac{2h+1}{2N}, \alpha\right)$, to be more precise) for any non-negative integer $l$, we may deduce the corresponding formula for the Dirichlet $L$-function $L(s, \chi)$, $\chi \bmod q$, at $s = -l + \frac{2h+1}{2N}$ via the well-known formula

$$q^{-s} \sum_{r=1}^{q} \chi(r)\zeta\left(s, \frac{r}{q}\right) = L(s, \chi) \tag{1.1}$$

and also at any rational point of the form $\frac{2h+1}{2N}$ through the functional equation. The procedure is the same as M. Katsurada's [3] who deduces his Corollary 2.1, stating the

The authors are supported by Grant-in-Aid for Scientific Research No. 14540051, 14540021 and 14005245 respectively.

closed form of $L\left(\frac{1}{2}, \chi\right)$ due to S. Egami [1], from Corollary 1.1 [3] on the value of $\zeta\left(\frac{1}{2}, \alpha\right)$.

It should be noted that Katsurada mentions that his Corollary 2.2 is a consequence of Corollary 2.1, but in the proof on p. 171 of Corollary 2.2, he deduces it from Corollary 1.1.

Thus our aim in this paper is to prove a theorem giving a closed formula for $L\left(-l + \frac{2h+1}{2N}, \chi\right)$ (and *a fortiori* for $L\left(\frac{2h+1}{2N}, \chi\right)$) and give some numerical examples.

**Notation.** We use the same notation similar to that in [2].

$N$ is a given positive integer, $h$ and $l$ are non-negative integers, $h < N$.

$$\begin{aligned} a_j &= \cos\left(\frac{\pi}{2N}\left(\frac{1}{2} - j\right)\right), \\ b_j &= \sin\left(\frac{\pi}{2N}\left(\frac{1}{2} - j\right)\right), \\ B_j(n, h, l, \alpha) &= -2\pi\alpha n - \frac{2h+1}{2N}\pi\left(\frac{1}{2} - j\right) - \frac{\pi(l-1)}{2}, \\ A(y) &= \pi(2\pi y)^{\frac{1}{2N}}, \end{aligned} \tag{1.2}$$

and

$$\begin{aligned} f_j(n, h, l, x, \alpha) &= \frac{\cos\left(2A\left(\frac{n}{x}\right)b_j + B_j(n, h, l, \alpha)\right) - e^{-2A\left(\frac{n}{x}\right)a_j}\cos(B_j(n, h, l, \alpha))}{\cosh\left(2A\left(\frac{n}{x}\right)a_j\right) - \cos\left(2A\left(\frac{n}{x}\right)b_j\right)} \\ g_j(n, h, l, x, \alpha) &= \frac{\sin\left(2A\left(\frac{n}{x}\right)b_j + B_j(n, h, l, \alpha)\right) - e^{-2A\left(\frac{n}{x}\right)a_j}\sin(B_j(n, h, l, \alpha))}{\cosh\left(2A\left(\frac{n}{x}\right)a_j\right) - \cos\left(2A\left(\frac{n}{x}\right)b_j\right)} \end{aligned} \tag{1.3}$$

($g_j$ is obtained by replacing cosines by sines in the numerator of $f_j$).

For a Dirichlet character $\chi$ mod $q$, denote the Gauss sum by

$$\tau(\chi) = \sum_{\substack{r=1 \\ (r,q)=1}}^{q} \chi(r) e\left(\frac{r}{q}\right), \tag{1.4}$$

where $e(x) = e^{2\pi i x}$.

**Theorem.** *Let $\chi$ be a primitive Dirichlet character* mod $q$, $q \geq 3$ *and let $L(s, \chi)$ be the associated Dirichlet L-function. Then under the above notation we have*

$$\sum_{r=1}^{q}\sum_{n=1}^{\infty} \frac{\chi(r)}{n^{2Nl-2h}} \frac{\exp\left(-\frac{r}{q}n^{2N}x\right)}{1 - \exp\left(-n^{2N}x\right)}$$

$$= \frac{(-1)^h}{2N}\left(\frac{2\pi}{x}\right)^{-l+\frac{2h+1}{2N}} \tau(\chi) \sum_{j=-(N-1)}^{N-1}{}''$$

$$\times \left\{ \frac{1+\chi(-1)}{2} \sum_{n=1}^{\infty} \overline{\chi}(n) f_j(n,h,l,x,0) n^{-1-l+\frac{2h+1}{2N}} \right.$$
$$\left. + \frac{1-\chi(-1)}{2i} \sum_{n=1}^{\infty} \overline{\chi}(n) g_j(n,h,l,x,0) n^{-1-l+\frac{2h+1}{2N}} \right\}$$
$$+ \sum_{j=0}^{l} \frac{(-1)^j}{j!} L(-j,\chi)\zeta(2N(l-j)-2h)\left(\frac{x}{q}\right)^j$$
$$+ \frac{1}{2N}\Gamma\left(-l+\frac{2h+1}{2N}\right) L\left(-l+\frac{2h+1}{2N},\chi\right)\left(\frac{x}{q}\right)^{l-\frac{2h+1}{2N}}, \tag{1.5}$$

*where* $\sum''^{N-1}_{j=-(N-1)}$ *means the summation over* $j=-(N-1), -(N-3), \ldots, N-3, N-1$.

The case $l=0$ gives

**Corollary 1.1.**

$$\sum_{r=1}^{q}\sum_{n=1}^{\infty} \frac{\chi(r) n^{2h} \exp\left(-\frac{r}{q} n^{2N} x\right)}{1-\exp\left(-n^{2N}x\right)}$$
$$= \frac{(-1)^h}{2N}\left(\frac{2\pi}{x}\right)^{\frac{2h+1}{2N}} \tau(\chi) \sum_{j=-(N-1)}^{N-1}{}''$$
$$\times \left\{ \frac{1+\chi(-1)}{2} \sum_{n=1}^{\infty} \overline{\chi}(n) f_j(n,h,0,x,0) n^{-1+\frac{2h+1}{2N}} \right.$$
$$\left. + \frac{1-\chi(-1)}{2i} \sum_{n=1}^{\infty} \overline{\chi}(n) g_j(n,h,0,x,0) n^{-1+\frac{2h+1}{2N}} \right\}$$
$$+ L(0,\chi)\zeta(-2h) + \frac{1}{2N}\Gamma\left(\frac{2h+1}{2N}\right) L\left(\frac{2h+1}{2N},\chi\right)\left(\frac{x}{q}\right)^{-\frac{2h+1}{2N}}, \tag{1.6}$$

where

$$f_j(n,h,0,x,0)$$
$$= \frac{\sinh\left(2A\left(\frac{n}{x}\right)a_j\right)\cos\left(B_j(n,h,0,0)\right) - \sin\left(2A\left(\frac{n}{x}\right)b_j\right)\sin\left(B_j(n,h,0,0)\right)}{\cosh\left(2A\left(\frac{n}{x}\right)a_j\right) - \cos\left(2A\left(\frac{n}{x}\right)b_j\right)}$$
$$-\cos\left(B_j(n,h,0,0)\right),$$

$$g_j(n,h,0,x,0)$$
$$= \frac{\sinh\left(2A\left(\frac{n}{x}\right)a_j\right)\sin\left(B_j(n,h,0,0)\right) + \sin\left(2A\left(\frac{n}{x}\right)b_j\right)\cos\left(B_j(n,h,0,0)\right)}{\cosh\left(2A\left(\frac{n}{x}\right)a_j\right) - \cos\left(2A\left(\frac{n}{x}\right)b_j\right)}$$
$$-\sin\left(B_j(n,h,0,0)\right). \tag{1.7}$$

In particular, for $N = 1, h = 0$, Corollary 1.1 reduces to

$$\sum_{r=1}^{q}\sum_{n=1}^{\infty}\frac{\chi(r)\exp\left(-\frac{r}{q}n^2x\right)}{1-\exp\left(-n^2x\right)}$$

$$= \frac{1}{2}\sqrt{\frac{2\pi}{x}}\tau(\chi)\left\{\frac{1+\chi(-1)}{2}\sum_{n=1}^{\infty}\frac{\overline{\chi}(n)}{\sqrt{n}}\frac{1}{\sqrt{2}}\left(\frac{\sinh\left(2\pi\sqrt{\frac{\pi n}{x}}\right)-\sin\left(2\pi\sqrt{\frac{\pi n}{x}}\right)}{\cosh\left(2\pi\sqrt{\frac{\pi n}{x}}\right)-\cos\left(2\pi\sqrt{\frac{\pi n}{x}}\right)}-1\right)\right.$$

$$\left.+\frac{1-\chi(-1)}{2i}\sum_{n=1}^{\infty}\frac{\overline{\chi}(n)}{\sqrt{n}}\frac{1}{\sqrt{2}}\left(\frac{\sinh\left(2\pi\sqrt{\frac{\pi n}{x}}\right)+\sin\left(2\pi\sqrt{\frac{\pi n}{x}}\right)}{\cosh\left(2\pi\sqrt{\frac{\pi n}{x}}\right)-\cos\left(2\pi\sqrt{\frac{\pi n}{x}}\right)}-1\right)\right\}$$

$$-\frac{1}{2}L(0,\chi)+\frac{\sqrt{\pi}}{2}L\left(\frac{1}{2},\chi\right)\left(\frac{q}{x}\right)^{\frac{1}{2}}, \tag{1.8}$$

which is Egami's result ([1], Theorem) stated as Corollary 2.2 in Katsurada [3].

## 2. Proof

Our proof is a rather straightforward consequence of the following lemma, which is Theorem 2.1 in [2].

**Lemma 2.1.** *In the statement of §1, we have for $x > 0$*

$$\sum_{n=1}^{\infty}\frac{1}{n^{2Nl-2h}}\frac{\exp\left(-\alpha n^{2N}x\right)}{1-\exp\left(-n^{2N}x\right)}$$

$$= \frac{(-1)^h}{2N}\left(\frac{2\pi}{x}\right)^{-l+\frac{2h+1}{2N}}\sum_{j=-(N-1)}^{N-1}{}''\sum_{n=1}^{\infty}f_j(n,h,l,x,\alpha)n^{-1-l+\frac{2h+1}{2N}}+P(x), \tag{2.1}$$

*where*

$$P(x) = \zeta(2N(l+1)-2h)x^{-1}$$
$$+\sum_{j=0}^{l}\frac{(-1)^j}{j!}\zeta(-j,\alpha)\zeta(2N(l-j)-2h)x^j$$
$$+\frac{1}{2N}\Gamma\left(-l+\frac{2h+1}{2N}\right)\zeta\left(-l+\frac{2h+1}{2N},\alpha\right)x^{l-\frac{2h+1}{2N}} \tag{2.2}$$

*and $\sum''$ has the same meaning as Theorem.*

*Proof of Theorem.* In view of (1.1), the proof will be complete if we sum (2.1) with $\alpha = \frac{r}{q}$ over $1 \le r \le q$ after multiplying it by $\chi(r)$ and express the sum in closed form.

Since the sum of the left-hand side member times $\chi(r)$ gives the left-hand side member of (1.5), and the sum of the last two terms on the right of (1.5), it suffices to evaluate

$$\sum_{r=1}^{q} \chi(r) f_j\left(n, h, l, x, \frac{r}{q}\right).$$

To this end it is enough to evaluate the numerator of $f_j$. Thus denoting by $A$ either $2A\left(\frac{n}{x}\right) b_j$ or 0, we have

$$\sum_{r=1}^{q} \chi(r) \cos\left(B_j\left(n, h, l, \frac{r}{q}\right) + A\right)$$

$$= \sum_{r=1}^{q} \chi(r) \cos\left(2\pi n \frac{r}{q} + \frac{\pi(2h+1)}{2N}\left(\frac{1}{2} - j\right) + \frac{\pi(l-1)}{2} - A\right)$$

$$= \cos\left(\frac{\pi(2h+1)}{2N}\left(\frac{1}{2} - j\right) + \frac{\pi(l-1)}{2} - A\right) \mathcal{C}(q)$$

$$- \sin\left(\frac{\pi(2h+1)}{2N}\left(\frac{1}{2} - j\right) + \frac{\pi(l-1)}{2} - A\right) \mathcal{S}(q),$$

where

$$\mathcal{C}(q) = \sum_{r=1}^{q} \chi(r) \cos \frac{2\pi n r}{q}$$

and

$$\mathcal{S}(q) = \sum_{r=1}^{q} \chi(r) \sin \frac{2\pi n r}{q}.$$

In view of $\mathcal{C}(q) = \frac{1+\chi(-1)}{2} \overline{\chi}(n) \tau(\chi)$ and $\mathcal{S}(q) = \frac{1-\chi(-1)}{2i} \overline{\chi}(n) \tau(\chi)$, we conclude that

$$\sum_{r=1}^{q} \chi(r) f_j\left(n, h, l, x, \frac{r}{q}\right) = \frac{1+\chi(-1)}{2} \overline{\chi}(n) \tau(\chi) f_j(n, h, l, x, 0)$$

$$+ \frac{1-\chi(-1)}{2i} \overline{\chi}(n) \tau(\chi) g_j(n, h, l, x, 0), \tag{2.3}$$

and that

$$\sum_{r=1}^{q} \chi(r) \cos\left(B_j\left(n, h, l, \frac{r}{q}\right) + A\right)$$

$$= \frac{1+\chi(-1)}{2} \overline{\chi}(n) \tau(\chi) \cos\left(-\frac{\pi(2h+1)}{2N}\left(\frac{1}{2} - j\right) - \frac{\pi(l-1)}{2} + A\right)$$

$$+ \frac{1-\chi(-1)}{2i} \overline{\chi}(n) \tau(\chi) \sin\left(-\frac{\pi(2h+1)}{2N}\left(\frac{1}{2} - j\right) - \frac{\pi(l-1)}{2} + A\right). \tag{2.4}$$

□

## 3. Numerical examples

In this section we shall give two numerical examples $L\left(\frac{1}{2}, \chi_4\right)$ and $L\left(\frac{1}{4}, \chi_4\right)$, where $\chi_4$ is the primitive Dirichlet character mod 4. In this case, we have $\chi_4(-1) = -1$ and $L(0, \chi_4) = \frac{1}{2}$. We shall choose $0 < x \leq 1$.

Firstly we take $N = 1$ and $h = 0$ in Corollary 1.1 (see also (1.8)). Then we have

$$\sum_{n=1}^{\infty} \frac{\exp\left(-\frac{1}{4}n^2 x\right)}{1+\exp\left(-\frac{1}{2}n^2 x\right)} = \left(\frac{\pi}{x}\right)^{\frac{1}{2}} \sum_{n=1}^{\infty} \frac{\chi_4(n)}{\sqrt{n}} \frac{\sin\left(2\pi\sqrt{\frac{\pi n}{x}}\right) + \cos\left(2\pi\sqrt{\frac{\pi n}{x}}\right) - e^{-2\pi\sqrt{\frac{\pi n}{x}}}}{\cosh\left(2\pi\sqrt{\frac{\pi n}{x}}\right) - \cos\left(2\pi\sqrt{\frac{\pi n}{x}}\right)}$$
$$- \frac{1}{4} + \left(\frac{\pi}{x}\right)^{\frac{1}{2}} L\left(\frac{1}{2}, \chi_4\right). \tag{3.1}$$

We truncate the above sums at $M$ and $L$ ($M \geq 1$, $L \geq 1$). After some calculations, we have

$$L\left(\frac{1}{2}, \chi_4\right) = \left(\frac{x}{\pi}\right)^{\frac{1}{2}} \left(\sum_{n\leq M} \frac{\exp\left(-\frac{1}{4}n^2 x\right)}{1+\exp\left(-\frac{1}{2}n^2 x\right)} + \frac{1}{4}\right)$$
$$- \sum_{n\leq L} \frac{\chi_4(n)}{\sqrt{n}} \frac{\sin\left(2\pi\sqrt{\frac{\pi n}{x}}\right) + \cos\left(2\pi\sqrt{\frac{\pi n}{x}}\right) - e^{-2\pi\sqrt{\frac{\pi n}{x}}}}{\cosh\left(2\pi\sqrt{\frac{\pi n}{x}}\right) - \cos\left(2\pi\sqrt{\frac{\pi n}{x}}\right)}$$
$$+ E_1(x, M, L) \tag{3.2}$$

with

$$|E_1(x, M, L)| \leq \frac{2}{\sqrt{\pi}} \left(\frac{1}{\sqrt{x}M} \exp\left(-\frac{x}{4}M^2\right) + \frac{\sqrt{x}}{2} \exp\left(-2\pi\sqrt{\frac{\pi L}{x}}\right)\right).$$

Now we take $x = 0.2$, $M = 63$ and $L = 60$, then the above fomula gives us

$$L\left(\frac{1}{2}, \chi_4\right) = 0.66769145718960917665869092930024848225159 7$$
$$82974293709774979865732176160878928098610 1\ldots,$$

which is valid at least 80 digits.

Secondly we take $N = 2$ and $h = 0$. From Corollary 1.1, we have

$$\sum_{n=1}^{\infty} \frac{\exp\left(-\frac{1}{4}n^4 x\right)}{1+\exp\left(-\frac{1}{2}n^4 x\right)} = 2^{-\frac{3}{4}} \left(\frac{\pi}{x}\right)^{\frac{1}{4}} \sum_{n=1}^{\infty} \frac{\chi_4(n)}{n^{\frac{3}{4}}} (g_{-1}(n, x) + g_1(n, x))$$
$$- \frac{1}{4} + 2^{-\frac{3}{2}} x^{-\frac{1}{4}} \Gamma\left(\frac{1}{4}\right) L\left(\frac{1}{4}, \chi_4\right), \tag{3.3}$$

where

$$\begin{aligned} g_{-1}(n,x) &= g_{-1}(n,0,0,x,0) \\ &= \frac{\sin\left(2A\left(\frac{n}{x}\right)\cos\frac{\pi}{8}+\frac{\pi}{8}\right) - e^{-2A\left(\frac{n}{x}\right)\sin\frac{\pi}{8}}\sin\frac{\pi}{8}}{\cosh\left(2A\left(\frac{n}{x}\right)\sin\frac{\pi}{8}\right) - \cos\left(2A\left(\frac{n}{x}\right)\cos\frac{\pi}{8}\right)} \\ g_{1}(n,x) &= g_{1}(n,0,0,x,0) \\ &= \frac{\cos\left(2A\left(\frac{n}{x}\right)\sin\frac{\pi}{8}-\frac{\pi}{8}\right) - e^{-2A\left(\frac{n}{x}\right)\cos\frac{\pi}{8}}\cos\frac{\pi}{8}}{\cosh\left(2A\left(\frac{n}{x}\right)\cos\frac{\pi}{8}\right) - \cos\left(2A\left(\frac{n}{x}\right)\sin\frac{\pi}{8}\right)}. \end{aligned}$$

We again truncate the sums in (3.3) at $M$ and $L$, and get

$$\begin{aligned} \Gamma\left(\frac{1}{4}\right)L\left(\frac{1}{4},\chi_4\right) &= 2\sqrt{2}x^{\frac{1}{4}}\left\{\sum_{n\le M}\frac{\exp\left(-\frac{1}{4}n^4x\right)}{1+\exp\left(-\frac{1}{2}n^4x\right)}+\frac{1}{4}\right\} \\ &\quad - 2^{\frac{3}{4}}\pi^{\frac{1}{4}}\sum_{n\le L}\frac{\chi_4(n)}{n^{\frac{3}{4}}}\left(g_{-1}(n,x)+g_1(n,x)\right) \\ &\quad + E_2(x,M,L) \end{aligned} \tag{3.4}$$

with

$$\begin{aligned} |E_2(x,M,L)| \le 2^{\frac{3}{2}}&\left\{\frac{1}{x^{\frac{3}{4}}M^3}\exp\left(-\frac{x}{4}M^4\right)\right. \\ &\left.+\frac{7}{2}x^{\frac{1}{4}}\exp\left(-(2\pi)^{\frac{5}{4}}\left(\frac{L}{x}\right)^{\frac{1}{4}}\sin\frac{\pi}{8}\right)\right\}. \end{aligned}$$

This time we take $x = 0.00001$, $M = 93$ and $L = 65$. Then we get from (3.4) that

$$\begin{aligned} L\left(\frac{1}{4},\chi_4\right) &= 0.590723056442494731865959153511562059798367 4 \\ &\quad 172391144008277187659300583182066459609692 8\ldots, \end{aligned}$$

which is also valid at least 80 digits.

## References


[1] S. Egami, A $\chi$-analogue of a formula of Ramanujan for $\zeta(1/2)$, *Acta Arith.* **69** (1995) 189–191

[2] S. Kanemitsu, Y. Tanigawa and M. Yoshimoto, On multiple Hurwitz zeta-function values at rational arguments, *Acta Arith.* **107** (2003) 45–67

[3] M. Katsurada, On an asymptotic formula of Ramanujan for a certain theta-type series, *Acta Arith.* **97** (2001) 157–172

Proc. Int. Conf. – Number Theory No. 1, 2004, pp. 39–50.

# Some Results and Problems on Arithmetical Functions

I. Kátai

Eötvös Loránd Tudományegyetem, Department of Computer Algebra and Research Group of Applied, Number Theory of the Hungarian Academy of Sciences, H-1117 Budapest, Pázmány Péter sétány 1/C.
e-mail: katai@compalg.inf.elte.hu

*Dedicated to Professor M.V. Subbarao on his 80th anniversary*

## 1. Introduction

Let, as usual, $\mathbb{N}$, $\mathbb{Z}$, $\mathbb{Q}$, $\mathbb{R}$, $\mathbb{C}$ be the set of positive integers, integers, rational, real, or complex numbers, respectively. Let $\mathbb{Q}_x$, $\mathbb{R}_x$ be the multiplicative group of positive rationals, reals, respectively. Let $\mathcal{P}$ be the set of prime numbers.

For an arbitrary, additively written Abelian group $G$ let $\mathcal{A}_G$, resp. $\mathcal{A}_G^*$ denote the classes of additive, resp. completely additive functions. A function $f : \mathbb{N} \to G$ belongs to $\mathcal{A}_G$ if $f(mn) = f(m) + f(n)$ holds for each pair of coprime $m, n$, and it belongs to $\mathcal{A}_G^*$ if the above equation holds for all $m, n \in \mathbb{N}$. If $G$ is multiplicatively written, then we shall write $\mathcal{M}_G$, $\mathcal{M}_G^*$ instead of $\mathcal{A}_G$, $\mathcal{A}_G^*$ and the corresponding functions are called multiplicative, completely multiplicative.

If $G = \mathbb{R}$, then we shall write simply $\mathcal{A}$, $\mathcal{A}^*$ instead of $\mathcal{A}_\mathbb{R}$, $\mathcal{A}_\mathbb{R}^*$. Similarly, if $G = \mathbb{C}$, then we shall write $\mathcal{M}$, $\mathcal{M}^*$ instead of $\mathcal{M}_\mathbb{C}$, $\mathcal{M}_\mathbb{C}^*$.

If $f \in \mathcal{A}_G^*$, then its domain $\mathbb{N}$ can be extended to $\mathbb{Q}_x$ by

$$f\left(\frac{m}{n}\right) := f(n) - f(n),$$

and the functional equation

$$f(r_1 r_2) = f(r_1) + f(r_2)$$

remains valid for every $r_1, r_2 \in \mathbb{Q}_x$.

Assume that $G$ is a topological group, $f \in \mathcal{A}_G^*$, and that $f : \mathbb{Q}_x \to G$ is continuous at 1. Then, for each $\alpha \in \mathbb{R}_x$ there exists the limit

$$\lim_{r \to \alpha} f(r) =: \phi(\alpha),$$

$\phi$ is continuous everywhere in $\mathbb{R}_x$, furthermore $\phi(\alpha\beta) = \phi(\alpha) + \phi(\beta)$ valid for $\alpha, \beta \in \mathbb{R}_x$, i.e. $\phi$ is a continuous homomorphism of $\mathbb{R}_x$ into $G$.

On the other hand, if $\phi : \mathbb{R}_x \to G$ is a homomorphism, then its restriction to the domain $\mathbb{N}$ is a completely additive function.

The question, under what condition is true that $f \in \mathcal{A}_G$ is close to a continuous homomorphism, is very delicate, and important as well. A survey of the most important results was given in [1]. We shall discuss some related questions in section.

## 2. Characterization of $n^s$ as a multiplicative function

In a series of papers [[2], I-IV.] I considered functions $f \in \mathcal{M}$ under the conditions that $\Delta f(n) = f(n+1) - f(n)$ tends to zero in some sense. I could determine all those pairs $f, g \in \mathcal{M}^*$ for which

$$\sum_{n=1}^{\infty} \frac{1}{n} |g(n+k) - f(n)| < \infty \tag{2.1}$$

with some fixed $k \in \mathbb{N}$, holds. Namely I proved

**Theorem 1.** *If $f, g \in \mathcal{M}$, (2.1) holds with $k = 1$, then either*

$$\sum \frac{|f(n)|}{n} < \infty, \quad \sum \frac{|g(n)|}{n} < \infty \tag{2.2}$$

*or*

$$f(n) = g(n) = n^{\sigma+i\tau} \quad \sigma, \tau \in \mathbb{R},\ 0 \le \sigma < 1. \tag{2.3}$$

**Theorem 2.** *Let $f, g \in \mathcal{M}^*$, $k \geq 2$. Assume that (2.1) holds, furthermore that $f(n) = g(n) = 0$ if $(n, k) > 1$, and $f(n) \neq 0$, $g(n) \neq 0$ if $(n, k) = 1$.*

*Then, either (2.2) is satisfied, or there exist $F, G \in \mathcal{M}^*$ and $s \in \mathbb{C}$ with $Res < 1$ such that $f(n) = n^s F(n)$, $g(n) = n^s G(n)$, and*

$$G(n + k) = F(n) \quad (n \in \mathbb{N}). \tag{2.4}$$

An immediate consequence of Theorem 1 is that $\sum \frac{1}{n} |\lambda(n + 1) - \lambda(n)| = \infty$, where $\lambda$ is the Liouwille function.

Recently B.M. Phong and I [3] proved

**Theorem 3.** *Let $k \in \mathbb{N}$ be fixed. Assume that (2.4) is satisfied. Let*

$$S_F := \{n \mid F(n) \neq 0\},\ S_G = \{n \mid G(n) \neq 0\}.$$

*Then, either $S_F$ and $S_G$ are finite sets, or $S_F$ contains all the integers $n \in \mathbb{N}$ which are coprime to $k$.*

In 1984 I was giving a talk in a conference celebrating P. Erdös, in Oatacamund. The next assertion was proposed by myself and proved by Wirsing in 1984.

**Theorem 4.** *If $f \in \mathcal{M}$, $\Delta f(n) \to 0$ as $n \to \infty$, then either $f(n) \to 0$ as $n \to \infty$, or $f(n) = n^s$, $0 \le Res < 1$.*

Some years later it was proved independently by Tang and Shao, and published in a joint paper with E. Wirsing [4].

*Conjecture* 1. Let $f \in \mathcal{M}^*, |f(n)| = 1 \quad (n \in \mathbb{N})$, and

$$\mathcal{B}_k = \{\alpha_1, \dots, \alpha_k\} = \text{ limit points of } \{f(n+1)\overline{f}(n) | \, n \in \mathbb{N}\}.$$

Then $\mathcal{B}_k = S_k = \{w | \, w^k = 1\}$, and $f(n) = n^{i\tau} F(n)$, where $F(\mathbb{N}) = S_k$ and for every $w \in S_k$ there is an infinite sequence $n_\nu$ such that $F(n_\nu + 1)\overline{F}(n_\nu) = w$.

In our joint papers with M.V. Subbarao [5], [6] we proved

**Theorem 5.**

(1) *Conjecture 1 is true for* $k = 1, 2, 3$.

(2) *Let* $k = 4$, *and assume that the conditions of Conjecture 1 are satisfied. Then there is some* $\tau \in \mathbb{R}$ *such that* $f(n) = n^{i\tau} F(n)$ *and either (A) or (B) holds.*

   (A) $F(\mathbb{N})(= \mathcal{B}_4) = S_4$

   (B) $\mathcal{B}_4$ *consists of four distinct elements of* $S_5$, *i.e.* $\mathcal{B}_4 = \{\kappa^{l_1}, \kappa^{l_2}, \kappa^{l_3}, \kappa^{l_4}\}$, *where* $\kappa \neq 1$ *is any element of* $S_5$, *moreover* $F(\mathbb{N}) = S_5$, *and for every* $w \in \mathcal{B}_4$, $F(n+1)\overline{F}(n) = w$ *holds for infinitely many* $n$.

*Remark.* We think that the case (B) cannot hold, which would follow if we could prove that $\mathbb{F}(\mathbb{N}) = S_5$ implies that $F(n+1)\overline{F}(n) = w$ has infinitely many solutions for every $w \in S_5$.

A conjecture weaker than Conjecture 1 can be formulated as

*Conjecture* 2. Let $f \in \mathcal{M}^*$, $|f(n)| = 1$ $(n \in \mathbb{N})$, and let $\mathcal{B}_k = \{\alpha_1, \dots, \alpha_k\}$. Then there is a suitable $m$, such that $\alpha_\nu^m = 1$ $(\nu = 1, \dots, k)$, consequently $f(n) = n^{i\tau} F(n)$, $F^m(n) = 1$ $(n \in \mathbb{N})$.

*Remark.* Recently I could prove that Conjecture 2 is true, if $k = 5$.

Let $\mathcal{A}_k = \{\alpha_1, \dots, \alpha_k\}$ be a finite set of complex numbers, $|\alpha_\nu| = 1$ $(\nu = 1, \dots, k)$. Let furthermore $C(n)$ be that value from $\mathcal{A}_k$, for which

$$|h_f(n) - C(n)| = \min_{j=1,\dots,k} |h_f(n) - \alpha_j|. \tag{2.5}$$

*Conjecture* 3. Assume that $f(n) \in \mathcal{M}^*$, $|f(n)| = 1$ $(n \in \mathbb{N})$, and that

$$\sum_{n<\infty} \frac{|h_f(n) - C(n)|}{n} < \infty, \tag{2.6}$$

and for every $\alpha_j \in \mathcal{A}$,

$$\sum_{\substack{n=1 \\ C(n)=\alpha_j}}^{\infty} \frac{1}{n} = \infty. \tag{2.7}$$

Then $f(n) = n^{i\tau} F(n)$, $F(\mathbb{N}) = S_k = \mathcal{A}_k$ and for every $w \in \mathcal{A}_k$,

$$\sum_{\overline{F}(n+1)F(n)=w} \frac{1}{n} = \infty.$$

Let $\mathcal{A}_k = \{\alpha_1, \ldots, \alpha_k\}$ be a finite set of complex numbers, $|\alpha_\nu| = 1$ $(\nu = 1, \ldots, k)$, $f \in \mathcal{M}^*$, $|f(n)| = 1$ $(n \in \mathbb{N})$, $h_f(n) = \overline{f}(n+1)f(n)$, let $C(n)$ be defined as in (2.5).

*Conjecture* 4. Assume that

$$\frac{1}{x}\sum_{n\le x}|h_f(n) - C(n)| \to 0 \quad (n \to \infty), \tag{2.8}$$

and that

$$\overline{\lim}_{x\to\infty}\left(\frac{1}{x}\sum_{\substack{n\le x\\ C(n)=\alpha_j}} 1\right) > 0 \quad \text{for every} \quad j > 0.$$

Then $\mathcal{A}_k = S_k$, $f(n) = n^{i\tau}F(n)$, $F^k(n) = 1$ $(n \in \mathbb{N})$, and

$$\overline{\lim}_{x\to\infty}\frac{1}{x}\sum_{\substack{F(n+1)=wF(n)\\ n\le x}} 1 > 0.$$

## 3. Characterization of continuous homomorphisms as elements of $\mathcal{A}_G$ for compact groups

We investigated this topic in a series of papers written jointly with Z. Daróczy [5–11].

Let $f \in \mathcal{A}_G^*$, $S_f$ =the set of all limit points of $\{f(n) \mid n \in \mathbb{N}\}$. Then $S_f$ is a compact subgroup of $G$, $f(\mathbb{N}) \subseteq S_f$. Let $\Delta$ be the set of limit points of $\{f(n+1)-f(n) \mid n \in \mathbb{N}\}$.

*Conjecture* 5. We have $\Delta$ is a subgroup of $S_f$. There is a continuous homomorphism $\varphi : \mathbb{R}_x \to S_f$ such that $f(n) = \varphi(n) + F(n)$, and $F(\mathbb{N}) = \Delta$.

## 4. On additive functions mod 1

Let $T = \mathbb{R}/\mathbb{Z}$ be the one-dimensional torus. We say that $F \in \mathcal{A}_T$ is of finite support if $F(p^\alpha) = 0$ holds for every large prime $p$, and every $\alpha \in \mathbb{N}$. For $F_\nu \in \mathcal{A}_T$ $(\nu = 0, \ldots, k-1)$ let

$$L_n(F_0, \ldots, F_{k-1}) := F_0(n) + \ldots + F_{k-1}(n+k-1).$$

Let $\mathcal{L}_0^{(k)}$ be the space of those $k$-tuples $(F_0, F_1, \ldots, F_{k-1})$ of $F_\nu \in \mathcal{A}_T$ for which

$$L_n(F_0, \ldots, F_{k-1}) = 0 \quad (n \in \mathbb{N})$$

holds.

*Conjecture* 6. $\mathcal{L}_0^{(k)}$ is a finite dimensional $\mathbb{Z}$ module, and each $F_j$ is of finite support.

*Conjecture* 7. If $F_\nu \in \mathcal{A}_T$ ($\nu = 0, 1, \dots, k-1$), and

$$L_n(F_{0,\dots,}F_{k-1}) \to 0 \quad (n \to \infty),$$

then there exist suitable real numbers $\tau_0, \tau_1, \dots, \tau_{k-1}$ such that $\tau_0 + \dots + \tau_{k-1} = 0$, and for $H_j(n) := F_j(n) - \tau_j \log n$ ($j = 0, \dots, k-1$),

$$L_n(H_0, \dots, H_{k-1}) = 0 \quad (n = 1, 2, \dots).$$

*Remarks.*

1) Conjecture 7 for $k = 1$ can be deduced easily from Wirsing's theorem.
2) Conjecture 6 is proved for $k \le 3$ (see [12], [13])

Let $P(n)$ be the largest and $p(n)$ the smallest prime divisor of $n$.

*Conjecture* 8. For every integer $k(\ge 1)$ there exists a constant $c_k$ such that for every prime $p$ greater than $c_k$,

$$\min_{\substack{j \ge 1 \\ P(j) < p}} \max_{\substack{l \in [-k,k] \\ l \ne 0}} P(jp + l) < p. \tag{4.1}$$

*Remark.* We are unable to prove it even for $k = 2$.

**Proposition 1.** *Let $\mathcal{L}_0^{*(l)}$ be the space of those $l$-tuples $(F_0, \dots, F_{l-1})$ of $F_\nu \in \mathcal{A}_T^*$ for which $L_n(F_0, \dots, F_{l-1}) = 0$ ($n \in \mathbb{N}$). Assume that Conjecture 8 is true for $k = l$. Then $\mathcal{L}_0^{*(l)}$ is a finite dimensional space.*

Let $Q_x^l$ be the group of the $l$-fold direct product of $Q_x$.
Let furthermore $O_l$ be the subgroup in it, generated by the elements

$$(n, n+1, \dots, n+l-1) \quad n \in \mathbb{N}.$$

The following assertion is true.

**Proposition 2.** *$\mathcal{L}_0^{*(l)}$ (defined in Proposition 1) is of finite-dimensional, if and only if the factor group $Q_x^l / O_l$ is finite. $\mathcal{L}_0^{*(l)}$ is trivial (it contains only the vectorial zero) if and only if $O_l = Q_x^l$.*

## 5. On almost monotonic additive functions

Our paper [14] written jointly with M.V. Subbarao, contains the following results.

If $f \in \mathcal{A}^*$, $t \ge 1$, $i_1, i_2, \dots, i_t$ is an arbitrary permutation of the integers $1, 2, \dots, t$ then

$$\frac{1}{x}\#\{n \le x \mid f(n+i_1) \le f(n+i_2) \le \dots \le f(n+i_t)\} \tag{5.1}$$

has presumably a positive limit for $x \to \infty$, except some "exceptional" functions.

Since the log function is monotonic, therefore it is exceptional. Another exceptional function $f$ can be constructed by choosing $f(2) > 0$, $f(p) = 0$ for every $p \in \mathcal{P}\backslash\{2\}$. Then

$$f(n+1) \le f(n+2) \le f(n+3)$$

has no solutions.

*Conjecture* 9. Assume that for $f \in \mathcal{A}^*$ there exists an integer $t \ge 1$, and a permutation $i_1, i_2, \ldots, i_t$ of the integers $1, 2, \ldots, t$ such that (5.1) tends to zero as $x \to \infty$. Then $f(n) = c \log n + u(n)$, with some constant $c$, where $u \in \mathcal{A}^*$ is of finite support.

Let $\mathcal{F}$ be the class of those monotonically decreasing functions $l : \mathbb{N} \to [0, \infty)$ for which

$$\sum_{j=1}^{\infty} l(2^j) < \infty.$$

**Theorem 6.** *Let $u \in \mathcal{A}^*$ be such a function, for which there is a suitable $l \in \mathcal{F}$ such that*

$$u(n) \le \max\,(u(n+1), \ldots, u(n+k)) + l(n) \quad (n \in \mathbb{N}) \tag{5.2}$$

*$k$ is a fixed integer.*

*Then there is a constant $c$ and $v \in \mathcal{A}^*$ such that $u(n) = c \log n + v(n)$, where $v(p) = 0$ for all but finitely many primes. If $\mathcal{R} = \{q_1, \ldots, q_r\}$ (it might be empty) is the set of the exceptional primes on which $v$ does not equal to zero, then $v(q_j) < 0$ $(j = 1, \ldots, r)$ and for every $n \in \mathbb{N}$ there exists a $j \in \{1, \ldots, k\}$, for which $n + j$ is coprime to each $q_l$ $(l = 1, \ldots, r)$.*

**Theorem 7.** *Assume that $u \in \mathcal{A}^*$, $l \in \mathcal{F}$, and*

$$u(n) \ge \min(u(n-1),\ u(n+1)) - l(n) \tag{5.3}$$

*holds. Then $u(n) = c \log n + v(n)$, where $c$ is an arbitrary constant, and either $v(n) = 0$ identically or there is an odd prime $q$ on which $v(q) > 0$, and $v(n) = 0$ if $(n, q) = 1$.*

*Conversely, if $u$ satisfies the above conditions, then (5.3) must be hold with some $l \in \mathcal{F}$.*

By using such type of theorems, we deduced

**Theorem 8.** *Let $f \in \mathcal{M}^*$, such that $f(n)$ is positive for every $n \in \mathbb{N}$, and that*

$$2f(n) \le f(n+1) + f(n-1)$$

*holds for every large $n$. Then $f(n) = n^s$ and either $s \le 0$, or $s \ge 1$.*

*Let $g \in \mathcal{M}^*$, $g$ is positive for every $n$, such that*

$$2g(n) \ge g(n+1) + g(n-1)$$

*holds for every large $n$. Then $g(n) = n^s$, $0 \le s \le 1$.*

## 6. Integer-valued multiplicative functions with congruence conditions

6.1. In 1966 M.V. Subbarao proved [15]

**Theorem 9.** *If* $f \in \mathcal{M}$, $f$ *is integervalued, and*

$$f(n+m) \equiv f(m) \pmod{n} \quad \textit{for all } m, n \in \mathbb{N} \tag{6.1}$$

*then* $f(n) = n^a$, $a \in \mathbb{N}$.

This nice theorem has been extended by some Hungarian mathematicians (as A. Iványi, I. Joó, J. Fehér, B.M. Phong). I wanted to mention only

**Theorem 10 (B.M. Phong [16]).** *If* $A$, $B$ *are positive coprime integers,* $f$ *is an integer valued multiplicative function, and*

$$f(An+B) \equiv C \pmod{n}$$

*holds for every* $n > n_0$, *then for* $(n, A) = 1$ *we have* $f(n) = \chi(n)n^\alpha$, *where* $\alpha$ *is a nonnegative integer,* $\chi$ *is a Dirichlet character* mod $A$.

In 1985 Subbarao introduced the notion of quasi multiplicative functions according to: $f$ is a quasi-multiplicative, if for every prime $p$ and every $n$ coprime to $p$,

$$f(np) = f(n) \cdot f(p).$$

*Conjecture* 10. (due to Fabrykowski and Subbarao). Let $f$ be an integer valued quasi multiplicative function, and let $\mathcal{R}$ be a subset of primes such that

$$\begin{cases} f(n+p) \equiv f(n) \pmod{p} \\ \text{holds for every } n \in \mathbb{N}, \text{ and } p \in \mathcal{R}. \end{cases} \tag{6.2}$$

If $\#\mathcal{R} = \infty$, then $f(n) = n^\alpha$, $\alpha \in \{0, 1, 2, \dots\}$.

This conjecture seems to be very deep. A partial important result is proved by J. Fehér and B.M. Phong [17]), according to: if $f$ satisfies the conditions, and $\#(\mathcal{P} \backslash \mathcal{R}) < \infty$, then $f(n) = n^\alpha$, $\alpha \in \{0, 1, 2, \dots\}$.

6.2. Now we formulate another kind of generalization of the original problem of Subbarao.

**Theorem 11.** *Let* $u_1(n), \dots, u_k(n)$ *be distinct integer valued completely multiplicative functions,* $c_1, \dots, c_k \in \mathbb{Z}$,

$$S(n) := c_1 u_1(n) + \dots + c_k u_k(n).$$

*Assume that*

$$S(n+m) \equiv S(m) \pmod{n} \tag{6.3}$$

*holds for every* $n, m \in \mathbb{N}$.

*Then* $S(n)$ *is a polynomial in* $n$.

*Proof.* For each $p \in \mathcal{P}$, let $u_j(p) = p^{\gamma_j(p)} t_j(p)$, where $(t_j(p), p) = 1$. Let $a_j = \min_{p \in \mathcal{P}} \gamma_j(p)$. Then we can write $u_j(n) = l_j(n) n^{a_j}$, and so

$$S(n) = c_1 l_1(n) n^{a_1} + \ldots + c_k l_k(n) n^{a_k}.$$

We shall assume that $a_1 \geq \cdots \geq a_k$.

Now

$$0 (\bmod n) \equiv S(n+m) - S(m) = \sum_h c_h m^{a_h} (l_h(n+m) - l_h(n)) (\bmod n)$$

whence by the substitution $n = km$,

$$0 \bmod n \equiv S((K+1)m) - S(m) \equiv \sum_h c_h m^{a_h} l_h(m) \, (l_h(K+1) - 1) (\bmod Km). \tag{6.4}$$

Let $U = \{h \mid a_h = 0\}$.

*Case* 1. Assume that $U$ is non-empty.

From (6.4) we obtain that

$$\sum_{h \in U} l_h(m) d_h \equiv 0 (\bmod m), \tag{6.5}$$

where

$$d_h = c_h (l_h(K+1) - 1). \tag{6.6}$$

Let $h_1$ be an arbitrary element in $U$. Then there is a prime $p = p^*$, such that $p^* \nmid l_{h_1}(p^*)$. Let $V = \{h \in U, \ p^* \nmid l_h(p^*)\}$. Then $V$ is non-empty, $h_1 \in V$.

$$\sum_{h \in U \setminus V} l_h^{\nu}(p^*) d_h \equiv O (\bmod p^{*\nu})$$

and form (6.5) we obtain that

$$\sum_{h \in V} l_h^{\nu}(p^*) d_h \equiv 0 (\bmod p^{*\nu}). \tag{6.7}$$

Let $B = \sum_{h \in V} d_h$. We shall prove that $B = 0$. Indeed, let $H$ be so large that $p^{*H} > |B|$. Let $\nu \equiv O(\bmod \varphi(p^{*H}))$. Then $l_h^{\nu}(p^*) \equiv 1 \ (\bmod p^{*H})$, $\nu > H^*$, thus from (6.7) we obtain that

$$B = \sum d_h \equiv O \left(\bmod p^{*H}\right)$$

which implies that $B = 0$.

Thus

$$\sum_{h \in V} c_h (l_h(K+1) - 1) = 0. \tag{6.8}$$

To solve (6.8) we observe that in the case $\#V = 1$, $h_1 \in V$, $l_h(n) = 1$ is the only solution.

Furthermore (6.8) is true, if $\sum_{h\in V} c_h = 0$, and $l_{h_1}(n) = l_{h_2}(n)$ holds for every $n \in \mathbb{N}$. This case cannot be occur, since it implies that the representation is not the shortest, since $u_{h_1}(n) = l_{h_1}(n) = l_{h_2}(n) = u_{h_2}(n)$.

We shall prove that no other solutions exist. Let

$$T(n) = \sum_{h\in V} c_h l_h(n). \tag{6.9}$$

From (6.8) we have that $T(n) = T(1)$ for every $n \in \mathbb{N}$.

Let $V = \{h_1, \ldots, h_r\}$. We shall use induction on $r$.

Let $r = 2$. Then

$$T(1)(1 - l_1(m)) = T(mn) - l_1(m)T(n) = c_2(l_{h_2}(m) - l_{h_1}(m))l_{h_2}(n). \tag{6.10}$$

If $l_{h_2}(m) = l_{h_1}(m)$ holds for every $m$, then we are ready, the representation of $S(n)$ is not the shortest. If there is some $m = m_0$ for which $l_{h_2}(m_0) \neq l_{h_1}(m_0)$, then $l_{h_2}(n)$ =constant, $l_{h_2}(n) = 1$ $(n = 1, 2, \ldots)$. Consequently $c_1 l_{h_1}(n) = c_1$ for every $h$, i.e. $l_{h_1}(n) = 1$ identically.

Assume that the assertion is proved, if the number of summands is $\leq r - 1$.

Let $V = \{h_1, \ldots, h_r\}$. Then

$$T(1)(1 - l_1(m)) = T(mn) - l_1(m)T(n) = \sum_{j=2}^{r} c_h(l_{h_j}(m) - l_{h_1}(m))l_{h_j}(n),$$

and the value of the right hand side does not depend on $n$. We have that $l_{h_j}(n) = 1$ identically, if there is a suitable $m$ for which $l_{h_j}(m) \neq l_{h_1}(m)$. Such an $m$ always exists, since $l_{h_1}, l_{h_j}$ $(j \geq 2)$ are not the same.

If $h \in U$, then by choosing an appropriate prime $\tilde{p}$ such that $\tilde{p} \nmid l_h(\tilde{p})$, we arrive that $h \in V$, and so that $l_h(n) = 1$ identically.

Thus $U$ may contain only one element $h$, for which $l_h(n) = 1$ identically.

Thus we can redefine $S$ by $c_1u_1(n) + \ldots + c_{k-1}u_{k-1}(n)$ and observe that this satisfies the conditions of the theorem.

*Case* 2. Assume that $a_1 \geq a_2 \geq \cdots \geq a_k > 0$. Let $b_l = a_l - a_k$ $(l = 1, \ldots, k)$. Then $b_1 \geq b_2 \geq \cdots \geq b_k = 0$. From (6.4) we have

$$m^{a_k} \sum_{h=1}^{k} c_h l_h(m) m^{b_h} [l_h^{\nu}(p) - 1] \equiv 0 (\mathrm{mod}\ m(p^{\nu} - 1)).$$

Since $l_h(p^{\nu}) \equiv t_h(p^{\nu}) (\mathrm{mod}\ (p^{\nu} - 1))$, therefore

$$m^{a_k - 1} \sum_{h=1}^{k} c_h l_h(m) m^{b_h} [t_h^{\nu}(p) - 1] \equiv 0\ (\mathrm{mod}\ (p^{\nu} - 1)). \tag{6.11}$$

We know that $p \nmid t_h(p)$.

We would like to prove that $t_h(p) = 1$ for every $h = 1, \dots, k$ (and for every $p \in \mathcal{P}$).

Assume first that there is a prime $Q$, such that $Q \mid t_h(p)$ for some $h$. Let $M = \{h, Q \mid t_h(p)\}$. Choose a large $H$. It is clear that $Q \neq p$. If $\nu \equiv 0 \pmod{\varphi(Q^H)}$, then $Q^H$ divides the left hand side of (6.10) and so

$$m^{a_k-1} \sum_{h \in M} c_h l_h(m) m^{b_h} \equiv 0 \pmod{Q^H}.$$

Since $H$ is arbitrary large, therefore

$$\sum_{h \in M} c_h l_h(m) m^{b_h} = 0 \quad (m \in \mathbb{N}). \tag{6.12}$$

But no such relation is possible among multiplicative functions.

It remains the case when $t_h(p) = 1$ or $-1$, $R = \{h \mid t_h(p) = -1\}$, and $R$ is non-empty.

Let $\nu \to \infty$ over the odd numbers. Then from (6.11) we deduce that

$$\sum_{h \in R} c_h l_h(m) m^{b_h} = 0 \quad (m \in \mathbb{N}) \tag{6.13}$$

which is impossible.

We proved that for every prime $p$, and every $h$, $t_h(p) = 1$.

The theorem is proved.

## 7. On group valued $q$-additive functions

For some fixed integer $q \geq 2$ and $n \in \mathbb{N}_0$ let $\varepsilon_0(n), \varepsilon_1(n), \dots$ be the sequence of the digits in the $q$-ary expansion of $n$, i.e.

$$n = \varepsilon_0(n) + \varepsilon_1(n) q + \dots, \qquad \varepsilon_j(n) \in A_q = \{0, 1, \dots, q-1\}. \tag{7.1}$$

Let $G$ be an arbitrary Abelian group, and $A_q(G)$ be the set of $G$-valued $q$-additive functions. A function $\varphi : \mathbb{N}_0 \to G$ belongs to $A_q(G)$, if $\varphi(0) = 0$ and

$$\varphi(n) = \sum_{j=0}^{\infty} \varphi(\varepsilon_j(n) q^j) \quad (n \in \mathbb{N}_0). \tag{7.2}$$

The right hand side is a finite sum, since $\varepsilon_j(n) = 0$ if $q^j > n$.

We are interested in the value distribution of $\varphi$ when $n$ runs over some subset of the whole set of integers, like polynomial values, or polynomial values at prime places.

An old conjecture of Gelfond states that for $q = 2$, $\alpha(n) := \sum \varepsilon_j(m)$ ($\alpha$ is the sum of digits function)

$$\frac{1}{\pi(x)} \#\{p \leq x \mid \alpha(p) \equiv 0 (\mathrm{mod}\ 2)\} \to 1/2 \quad (x \to \infty)$$

where $p$ runs over $\mathcal{P}$. Presently it is not known, whether there do exist infinite subsequences of primes $\{p_j\}$, $\{q_j\}$, such that $\alpha(p_j)$ =even, $\alpha(q_j)$ =odd.

Let $G = \mathbb{R}/\mathbb{Z} = T$ the one-dimensional torns. It is quite plausible to guess that $\varphi \in \mathcal{A}_q(T)$ $\varphi(p) = 0$ $(p \in \mathcal{P})$ implies that $\varphi(nq) = 0$ $(n \in \mathbb{N})$.

In a joint paper with Indlekofer [18] we proved

**Theorem 12.** *There exist a numerical constant* $c_1$ *with the following property. If* $\varphi \in \mathcal{A}_q(T)$, $\varphi(p) = 0$ *for every large* $p$, *then* $k\varphi(nq) = 0$ *holds for every* $n \in \mathbb{N}_0$, *with a suitalbe* $k \in \mathbb{N}$ *from* $[1, c_1]$.

In a joint paper of M.V. Subbarao, and myself [19] we proved the following two theorems.

**Theorem 13.** *Let* $P(n) = A_k n^k + \ldots + A_1 n + A_0$, $A_j \in \mathbb{Z}$, $A_k > 0$. *Let* $G$ *be a topological Abelian group,* $f \in \mathcal{A}_q(G)$ *with the property that*

$$\lim_{n\to\infty} f(P(n)) = \alpha$$

*with some* $\alpha \in G$.

Then $f(P(n)) = f(P(0))$ $(n \in \mathbb{N}_0)$, and there exists an integer $v \in \mathbb{N}_0$ such that $f(nq^v) = nf(q^v)$ for $n \in \mathbb{N}_0$.

Let $\Delta$ be the greatest common divisor of the values $\{P(m) - A_0 \mid m \in \mathbb{N}_0\}$. Let $\Delta_1$ be the largest divisor of $\Delta$ which is coprime to $q$. Let $D_1$ be the largest divisor of $k!\ A_k$ which is coprime to $q$.

If $(D_1, \Delta_1) = 1$, then there exists an integer $M(\geq 0)$ such that $f(bq^u) = 0$ for every $b \in A_q$ and $u \geq M$.

**Theorem 14.** *Let* $q$ *be odd,* $q \geq 3$. *Let* $f \in \mathcal{A}_q(G)$, *and assume that the set* $\mathcal{E}$ *of the limit points of* $\{f(n^2) \mid n \in \mathbb{N}_0\}$ *is a finite set. Then* $Df(nq^R) = 0$ *with some integers* $D \geq 1$, $R \geq 0$.

## 8. Final remark

Conjecture 2 has been proved by E. Wirsing recently.

## References

[1] I. Kátai, *Continuous homomorphisms as arithmetical functions, and sets of uniqueness*, Number Theory, edited: R.P. Bambah, V.C. Dumir, R.J. Hans-Gill, Hindustian Book Agency and Indian National Science Academy, 2000, ISBN 81-85931-23-2, pp. 183–200

[2] I. Kátai, Multiplicative functions with regularity properties, I-IV *Acta Math. Hung.* **42** (1983), 295–308; **43** (1984) 105–130, 259–272, **44** (1984), 125–132; **45** (1985), 379–380; **58** (1991), 343–350

[3] I. Kátai and B.M. Phong, On some pairs of multiplicative functions correlated by an equation, New Trends in Probability and Statistics, *Analytic and Probabilistic Methods in Number Theory*, TEV, Vilnius, Lithuania **4**, 191–203

[4] E. Wirsing, Tang Yuansheng and Shao Pintsung, On a conjecture of Kátai for additive functions, *J. Number Theory* **56** (1996), 391–395

[5] Z. Daróczy and I. Kátai, On additive arithmetical functions with values in the circle group, *Publ. Math. Debrecen* **34** (1987), 307–312

[6] Z. Daróczy and I. Kátai, On additive arithmetical functions with values in topological groups, *Publ. Math. Debrecen* **33** (1986), 287–292

[7] Z. Daróczy and I. Kátai, On additive number-theoretical functions with values in a compact Abelian group, *Aequationes Math.* **28** (1985), 288–292

[8] Z. Daróczy and I. Kátai, On additive arithmetical functions with values in topological groups II, *Publ. Math. Debrecen* **34** (1987), 65–68

[9] Z. Daróczy and I. Kátai, On additive functions taking values from a compact group, *Acta Sci. Math.* **53** (1989), 59–65

[10] Z. Daróczy and I. Kátai, Characterization of additive functions with values in the circle group, *Publ. Math. Debrecen* **36** (1989), 1–7

[11] Z. Daróczy and I. Kátai, Characterization of pairs of additive functions with some regularity property, *Publ. Math. Debrecen* **37** (1990), 217–221

[12] I. Kátai, On additive functions satisfying a congruence, *Acta Sci. Math.* **47** (1984), 85–92

[13] R. Styer, A problem of Kátai on sums of additive functions, *Acta Sci. Math.* **55** (1991), 269–286

[14] I. Kátai and M. V. Subbarao, Local peaks of additive functions, *Publ. Math. Debrecen* (accepted)

[15] M. V. Subbarao, Arithmetic functions satisfying a congruence property, *Canad. Math. Bull.*, **9** (1966), 143–146

[16] B. M. Phong, Multiplicative functions satisfying a congruence property IV., *Acta Math. Hungar.* **62** (1993), 81–87

[17] J. Fehér and B. M Phong, On a problem of Fabrykowski and Subbarao concerning quasi multiplicative functions satisfying a congruence property, *Acta Math. Hungar.*, **89** (2000), 149–159

[18] K.-H. Indlekofer and I. Kátai, On $q$-multiplicative functions taking a fixed value on the set of primes, *Studia Math.* (accepted)

[19] I. Kátai and M. V Subbarao, Group valued $q$-additive functions on polynomial values, *Publ. Math. Debrecen* (accepted)

Proc. Int. Conf. – Number Theory No. 1, 2004, pp. 51–55.

# Some Results in Number Theory, II

V. Kumar Murty

Department of Mathematics, University of Toronto, 100 St. George Street, Toronto, ON M5S 1A1, Canada

*Dedicated to Professor M. V. Subbarao*

For a natural number $n$, denote by $\phi(n)$ the Euler phi function and by $\nu(n)$ the number of distinct prime factors of $n$. Let $k \geq 1$ be an integer and let us set

$$F_k(x) = \#\{n \leq x : \nu((n, \phi(n))) = k\}.$$

In part I of this paper, M. Ram Murty and the author [4] proved that

$$F_k(x) = \left(\frac{e^{-\gamma}}{k!} + \mathbf{o}(1)\right) \frac{x(\log\log\log\log x)^k}{\log\log\log x}.$$

The inspiration for this was an old result of Erdös [1] that proved this in the case $k = 0$.

The case $k = 0$ considered by Erdös has an interpretation in terms of enumeration of finite groups. Indeed, $F_0(x)$ is the number of integers $n \leq x$ for which there is only one isomorphism class of groups of order $n$. Let us set $G(n)$ to be the number of isomorphism classes of groups of order $n$ and

$$H_k(x) = \#\{n \leq x : G(n) = k\}.$$

Then, Erdös, Ram Murty and the author [2] proved that for $k = 2^\ell$, we have

$$H_k(x) = (e^{-\gamma} + \mathbf{o}(1)) \frac{x}{(\log\log\log x)^\ell}.$$

The problem of obtaining an asymptotic formula for $H_3(x)$ is still open. Several bounds are known.These and related results may be found in [2], [5], [6], [7]. The article by Ram Murty [3] gives a survey and overview of known results.

The purpose of this paper is to investigate a function which at first sight appears to be rather more complicated than $F_k(x)$. Consider the function

$$G_k(x) = \#\{n \leq x : (n, \phi(n)) = k\}.$$

Our aim is to prove the following result.

**Theorem 0.1.** *We have*

$$G_k(x) = \left(\frac{e^{-\gamma}}{k} + \mathbf{o}(1)\right) \frac{x}{\log\log\log x}.$$

We shall in fact show that this follows easily from the result of Erdös. It is somewhat surprising that the constant depends in such a simple manner on $k$. One might have expected that the arithmetic nature of $k$ should manifest itself in the constant. However, this dependence probably manifests itself in the error term and one would expect that making the dependence explicit may be a more difficult question. Also, note that the case $k = 1$ is the result of Erdös quoted above.

We can ask for many variants of this problem. For example, instead of distinct prime divisors,we can ask if the asymptotics change if we count prime divisors with multiplicity. Another direction is to replace the Euler function with other functions for which we know divisibility properties. Besides classical functions such as various divisor functions, examples arise from Fourier coefficients of cusp forms [8], [9].

Denoting by $a_n = a_n(f)$ the $n$-th Fourier coefficient of a normalized holomorphic Hecke eigenform $f$ of weight $\geq 2$ without complex multiplication, we expect that the congruence

$$a_p \equiv 0 \pmod{p}$$

should happen very rarely. A consequence of the methods that we develop here allows us to say something about a simpler quantity, namely $(n, a_n)$. In particular, it is possible to show that almost always, $a_n$ has a divisor $> 1$ in common with $n$. The details of this will be presented elsewhere.

## 1. Preliminaries

In order to prove the main result, we need two lemmas.

**Lemma 1.1.** *Let $\ell$ be a fixed prime. Then, the number of $m \leq x$ such that $m$ has no prime divisor $\equiv 1 \pmod{\ell}$ is $\mathbf{O}(x/(\log x)^{1/(\ell-1)})$.*

*Proof.* By Brun's sieve, this is

$$\ll x \prod_{\substack{q\equiv 1 \pmod{\ell} \\ q\leq x}} \left(1 - \frac{1}{q}\right).$$

This quantity is seen to be

$$\ll x \exp\left\{-\frac{\log \ell + \log\log x}{\ell - 1}\right\} = x(\log x)^{-1/(\ell-1)}.$$

**Lemma 1.2.** *If $\ell^a || k$, then*

$$\#\{n \leq x : \ell^{a+1}|n \text{ and } (n, \phi(n)) = k\} \ll \frac{x}{(\log x)^{1/\phi(\ell)}} = \mathbf{o}\left(\frac{x}{\log\log\log x}\right).$$

*Here, the implied constant depends on $\ell$.*

*Proof.* For an $n$ with $\ell^{a+1}|n$ we see that in fact $\ell^{a+1}||n$. Then, writing such an $n$ as $n = \ell^{a+1}m$, we have $\ell \not| m$ and

$$(\ell m, (\ell - 1)\phi(m)) = \frac{k}{\ell^a}.$$

Thus, $\ell \not| \phi(m)$, or what is the same, $m$ is not divisible by a prime $q \equiv 1 (\mathrm{mod}\, \ell)$. The number of such $m \leq x$ is by Lemma 1.1

$$\ll \frac{x}{(\log x)^{1/(\ell-1)}}.$$

Note that we may regard $k$ (and hence $\ell$) as fixed as $x \to \infty$. Thus, this quantity is

$$\mathbf{o}\left(\frac{x}{\log\log\log x}\right)$$

proving the lemma.

Let us set

$$L_1 = L_1(x) = \log x$$

and for each $i \geq 2$, define

$$L_i = L_i(x) = \log L_{i-1}(x).$$

To prove Theorem (0.1), we may assume that $n = km$ and $(k, m) = 1$. Indeed, if $\ell^a||k$ and $(n, \phi(n)) = k$, then certainly $\ell^a|n$. Lemma 1.2 shows that apart from a negligible number, $\ell^a||n$. This applies to each prime divisor $\ell$ of $k$. Thus $(k, n/k) = 1$.

Before proving the theorem, we need to recall a crucial result of Erdös [1].

**Lemma 1.3.** *The number of integers $m \leq x$ with the property that $(m, \phi(m)) = 1$ and having a prime divisor $< L_2(x)^{1+\epsilon}$ is*

$$\mathbf{o}(x/L_3(x)).$$

To prove this, Erdös uses Brun's sieve to show that the number of integers $m \leq x$ with $(m, \phi(m) = 1$ and having a prime divisor $< L_2(x)^{1-\epsilon}$ is $\mathbf{o}(x/L_3(x))$. He then shows that for a prime $p$ in the interval $(L_2(x)^{1-\epsilon}, L_2(x)^{1+\epsilon})$, the number of integers $m \leq x$ that are divisible by $p$ and that have all their prime divisors $\geq p$ is

$$\ll \sum_p \frac{x}{p \log p} \ll \frac{x}{L_3(x)} \sum_p \frac{1}{p}$$

and this is

$$\mathbf{o}(x/L_3(x)).$$

For our application, it suffices to know that apart from $\mathbf{o}(x/L_3(x))$ of the $m \leq x$, we may assume that all prime divisors of $m$ are larger than $g(x)$ for some function $g$ of $x$ with $g(x) \longrightarrow \infty$ as $x \longrightarrow \infty$.

## 2. Proof of the theorem

We have seen that

$$G_k(x) \sim \#\{n \le x : (k, n/k) = 1, \text{ and } (n, \phi(n)) = k\}.$$

Let us write $k = k_1 k_2$ where

$$k_1 = \prod_{\ell | k} \ell$$

and

$$k_2 = k/k_1.$$

Then, writing $n = k_1 k_2 m$ and assuming that $(m, k) = 1$, we see that

$$(n, \phi(n)) = (k_1 k_2 m, \phi(k_1) k_2 \phi(m)) = k_2 (k_1 m, \phi(k_1 m)).$$

Thus,

$$G_k(x) \sim G_{k_1}(x/k_2).$$

It therefore suffices to prove the theorem for $k$ squarefree. Then, the condition $(km, \phi(km)) = k$ is equivalent to three conditions:

(1) $(m, \phi(m)) = 1$

(2) $k | \phi(k)\phi(m)$

(3) $(m, \phi(k)) = 1$

Hence, we must count the number of integers $m \le x/k$ satisfying (1), (2) and (3). Equivalently, we must show that the number of $m \le x$ satisfying these three conditions is

$$(1 + \mathbf{o}(1)) \frac{x e^{-\gamma}}{\log \log \log x}.$$

By the theorem of Erdoš, we know that the number of $m \le x$ satisfying (1) has this asymptotic growth. It therefore remains to show that the number of $m \le x$ satisfying (1) and failing to satisfy (2) or (3) is $\mathbf{o}(x/L_3(x))$. Erdös' argument (see Lemma 1.3) shows that the number of $m \le x$ satisfying (1) and which have a prime divisor $< L_2(x)^{1+\epsilon}$ is $\mathbf{o}(x/L_3(x))$. Hence, we may assume that the $m \le x$ under consideration have all prime divisors $> L_2(x)^{1+\epsilon}$.

The number which have a prime divisor in common with $\phi(k)$ is therefore bounded as all prime divisors of $m$ are $> (\log \log x)^{1+\epsilon}$. Also, for each $\ell | k$, the number for which $\ell \nmid \phi(m)$ is (by Lemma 1.1), bounded by

$$\ll \frac{x}{(\log x)^{1/\phi(\ell)}} = \mathbf{o}\left( \frac{x}{L_3(x)} \right).$$

Hence, these are a negligible number as well. This proves the theorem.

## References


[1] P. Erdös, Some asymptotic formulas in number theory, *J. Indian Math. Soc.*, **12** (1949) 75–78

[2] P. Erdös, M. Ram Murty and V. Kumar Murty, On the enumeration of finite groups, *J. Number Theory* **25** (1987) 360–378

[3] M. Ram Murty, Counting finite groups, in: *Analysis, Geometry and Probability*, edited by Rajendra Bhatia, Hindustan Book Agency, (1996) 161–172

[4] M. Ram Murty and V. Kumar Murty, Some results in number theory I, *Acta Arith.*, **35** (1979) 367–371

[5] M. Ram Murty and V. Kumar Murty, On the density of various classes of groups, *J. Number Theory*, **17** (1983) 29–36

[6] M. Ram Murty and V. Kumar Murty, On the number of groups of a given order, *J. Number Theory*, **18** (1984) 179–191

[7] M. Ram Murty and V. Kumar Murty, On groups of squarefree order, *Math. Annalen* **267** (1984) 299–309

[8] M. Ram Murty and V. Kumar Murty, Prime divisors of Fourier coefficients of modular forms, *Duke Math. J.* **51** (1984) 57–77

[9] M. Ram Murty and V. Kumar Murty, An analogue of the Erds-Kac theorem for Fourier coefficients of modular forms, *Ind. J. Pure and App. Math.*, **15** (1984) 1090–1101

[10] V. Kumar Murty, A variant of Lehmer's conjecture, submitted 2004.

Proc. Int. Conf. – Number Theory No. 1, 2004, pp. 57–70.

# A New Theorem on Partitions

Padmavathamma, M. Ruby Salestina and S. R. Sudarshan

Department of Studies in Mathematics, University of Mysore, Manasagangotri, Mysore 570 006, India
e-mail: padma_vathamma@yahoo.com salestina@rediffmail.com
sudarshansr@softhome.net

**Abstract.** G. E. Andrews [A2] in 1974 conjectured and proved in 1994 [ABO] a result on two three-parameter partition functions $A_{4,3,3}(n)$ and $B^0_{4,3,3}(n)$. The object of this paper is to prove the new result $A_{5,3,3}(n) = B^0_{5,3,3}(n)$.

**Keywords.** Andrews' theorem, three parameters, partition functions.

**2000 AMS Subject Classification:** Primary 11P82, Secondary 05A19

## 1. Introduction

First we define the two three-parameter partition functions $A_{\lambda,k,a}(n)$ and $B_{\lambda,k,a}(n)$.

**Definition 1.1.** For an even integer $\lambda$, let $A_{\lambda,k,a}(n)$ denote the number of partitions of $n$ into parts such that no part $\not\equiv 0 (\text{mod } \lambda + 1)$ may be repeated and no part is $\equiv 0, \pm\left(a - \frac{\lambda}{2}\right)(\lambda+1)\,[\text{mod } (2k-\lambda+1)(\lambda+1)]$. For an odd integer $\lambda$, let $A_{\lambda,k,a}(n)$ denote the number of partitions of $n$ into parts such that no part $\not\equiv 0 \left(\text{mod } \frac{\lambda+1}{2}\right)$ may be repeated, no part is $\equiv \lambda + 1 (\text{mod } 2\lambda + 2)$ and no part is $\equiv 0, \pm(2a-\lambda)\left(\frac{\lambda+1}{2}\right)$ [mod $(2k-\lambda+1)(\lambda+1)$].

**Definition 1.2.** Let $B_{\lambda,k,a}(n)$ denote the number of partitions of $n$ of the form $b_1 + \cdots + b_s$ with $b_i \geq b_{i+1}$, no part $\not\equiv 0(\text{mod } \lambda + 1)$ is repeated, $b_i - b_{i+k-1} \geq \lambda + 1$ with strict inequality if $\lambda + 1/b_i$ and $\sum_{i=j}^{\lambda-j+1} f_i \leq a - j$ for $1 \leq j \leq \frac{\lambda+1}{2}$ and $f_1 + \ldots + f_{\lambda+1} \leq a - 1$ where $f_i$ is the number of appearances of $j$ in the partition.

Andrews [A1] proved the following theorem.

**Theorem 1 ([A1, Th. 2]).** *If $\lambda$, $k$, and $a$ are positive integers with $\frac{\lambda}{2} \leq a \leq k, k \geq 2\lambda - 1$ then for every positive integer $n$, we have $A_{\lambda,k,a}(n) = B_{\lambda,k,a}(n)$.*

For values $k < \lambda$ in theorem 1, Schur [S] has proved that

$$A_{3,2,2}(n) = B^0_{3,2,2}(n)$$

where $B^0_{3,2,2}(n)$ is the number of partitions enumerated by $B_{3,2,2}(n)$ with the additional condition that no part is $\equiv 2(\text{mod } 4)$.

Andrews [A2] conjectured the following:

*Conjecture* 1 *(A3, 84–85).* For all $n \geq 0$, we have

$$A_{4,3,3}(n) = B^0_{4,3,3}(n)$$

where $B^0_{4,3,3}(n)$ is the number of partitions enumerated by $B_{4,3,3}(n)$ with the added conditions:

$$f_{5j+2} + f_{5j+3} \leq 1 \text{ for } j \geq 0$$

$$f_{5j+4} + f_{5j+6} \leq 1 \text{ for } j \geq 0$$

$$f_{5j-1} + f_{5j} + f_{5j+5} + f_{5j+6} \leq 3 \text{ for } j \geq 1$$

In 1994, Andrews, Bessenrodt and Olsson [ABO], proved this conjecture analytically. The first two authors of this paper, have proved the same result [PR] combinatorially.

In section 2 of this paper we prove combinatorially the next natural case when $k < \lambda$ which is stated below. We include a table of values of $A_{5,3,3}(n)$ for $n$ equal to 1–60 in section 3.

**Theorem 2.** *For all* $n \geq 0$, *we have*

$$A_{5,3,3}(n) = B^0_{5,3,3}(n)$$

*where* $B^0_{5,3,3}(n)$ *is the number of partitions enumerated by* $B_{5,3,3}(n)$ *with the added conditions:*

$$\text{no part is } \equiv 3 (\text{mod } 6)$$

$$f_{6j+2} + f_{6j+4} \leq 1 \text{ for } j \geq 0$$

$$f_{6j+5} + f_{6j+7} \leq 1 \text{ for } j \geq 0$$

$$f_{6j-1} + f_{6j} + f_{6j+6} + f_{6j+7} \leq 3 \text{ for } j \geq 1$$

## 2. Proof of theorem 2

We first construct the mapping from the partitions enumerated by $A_{5,3,3}(n)$ to those enumerated by $B^0_{5,3,3}(n)$. We observe that $A_{5,3,3,}(n)$ is the number of partitions of $n$ into distinct parts without multiples of 3. Let $\pi = b_1 + \cdots + b_s$ be a partition enumerated by $A_{5,3,3}(n)$. Let **DC** denote the difference condition "$b_i - b_{i+2} \geq 6$ with strict inequality if $b_i$ is a multiple of 6". First we note that if there is no 7–5 or 4–2 pair and if DC is satisfied for all the parts of $\pi$ then it is a partition enumerated by $B^0_{5,3,3}(n)$ also. We adopt the following procedure to go from $A$ to $B^0$.

**Step** $AB_1$**.** List the parts of $\pi$ in a column in decreasing order. Let $\pi^1$ denote this partition.

**Step $AB_2$.** In $\pi^1$ starting from the top we look for the quadruples of the type

$$\begin{pmatrix} x \\ 6j+2 \\ 6j+1 \\ 6j-1 \\ 6j-2 \\ y \end{pmatrix}, \ j \geq 1 \tag{1}$$

and

$$\begin{pmatrix} x \\ 6j+5 \\ 6j+4 \\ 6j+2 \\ 6j+1 \\ y \end{pmatrix}, \ j \geq 0 \tag{2}$$

where $(x, 6j+2)$ [resp. $(x, 6j+5)$] and $(6j-2, y)$ [resp. $(6j+1, y)$] are not 7–5 or 4–2 pairs (either of them can be a 7–5 or 4–2 pair but not both of them). Then we replace (1) and (2) by

$$\begin{pmatrix} x \\ 12j \\ 12j \\ y \end{pmatrix} \quad \text{and} \quad \begin{pmatrix} x \\ 12j+6 \\ 12j+6 \\ y \end{pmatrix}$$

respectively.

**eg:** i) $\begin{matrix} 11 \\ 10 \\ 8 \\ 7 \\ 5 \end{matrix} \to \begin{matrix} 18 \\ 18 \\ 5 \end{matrix}$ ii) $\begin{matrix} 19 \\ 8 \\ 7 \\ 5 \\ 4 \\ 1 \end{matrix} \to \begin{matrix} 19 \\ 12 \\ 12 \\ 1 \end{matrix}$

If both $(x, 6j+2)$ [resp. $(x, 6j+5)$] and $(6j-2, y)$ [resp. $(6j+1, y)$] are 7–5 or 4–2 pairs then we replace (1) and (2) by

$$\begin{pmatrix} x+(6j+2) \\ 12j \\ (6j-2)+y \end{pmatrix} \quad \text{and} \quad \begin{pmatrix} x+(6j+5) \\ 12j+6 \\ (6j+1)+y \end{pmatrix}$$

respectively i.e., we just add the pairs.

$$\textbf{eg:}\quad \text{i)}\quad \begin{array}{c}13\\11\\10\\8\\7\\5\\2\end{array} \rightarrow \begin{array}{c}24\\18\\12\\2\end{array} \qquad \text{ii)}\quad \begin{array}{c}10\\8\\7\\5\\4\\2\\1\end{array} \rightarrow \begin{array}{c}18\\12\\6\\1\end{array}$$

Let the resulting partition be $\pi^2$.

If parts of the type mentioned in step $AB_2$ are not there then we move directly from step $AB_1$ to step $AB_3$.

**Step** $AB_3$**.** Starting from the **top** add all the 7–5 and 4–2 pairs. Let $\pi^3$ denote the resulting partition. Suppose DC is satisfied for all the parts of $\pi^3$. Then rearrange the parts of $\pi^3$ in decreasing order. The resulting partition will be the required partition enumerated by $B^0_{5,3,3}(n)$.

*Note* 1. If steps $AB_2$ and $AB_3$ are not applicable to a given partition then we move directly from step $AB_1$ to step $AB_4$. While attending to DC, we do not do it simultaneously. Only after attending to DC for the first triple from the **top** (in one or more steps), we move to the next triple with difference $< 6$.

**Step** $AB_4$**.** If DC is not satisfied then we look for the first $i$ (from the top) for which there is a triplet

$$\begin{pmatrix} b_i \\ b_{i+1} \\ b_{i+2} \end{pmatrix} \tag{3}$$

with $b_i - b_{i+2} < 6$. Now we get the following possibilities:

*Case* 1. Let $b_i, b_{i+2} \not\equiv 0(\text{mod } 6)$. In this case $b_i > b_{i+2}$ and $b_i + b_{i+2} \equiv 0(\text{mod } 6)$. Here $b_{i+1}$ may or may not be a multiple of 6. We replace (3) by $\begin{pmatrix} b_i + b_{i+2} \\ b_{i+1} \end{pmatrix}$.

$$\textbf{eg.}\quad \text{i)}\quad \begin{array}{c}11\\10\\7\end{array} \rightarrow \begin{array}{c}18\\10\end{array} \qquad \text{ii)}\quad \begin{array}{c}17\\14\\13\end{array} \rightarrow \begin{array}{c}30\\14\end{array}$$

*Case* 2. Let $b_i, b_{i+1} \not\equiv 0(\text{ mod } 6)$ and $b_{i+2} \equiv 0(\text{ mod } 6)$. Suppose $b_i - 6 < b_{i+1}$. Then we replace (3) by $\begin{pmatrix} b_{i+2} + 6 \\ b_{i+1} \\ b_i - 6 \end{pmatrix}$.

$$\textbf{eg.}\quad \begin{array}{c}17\\14\\7\\5\\2\end{array} \rightarrow \begin{array}{c}\left.\begin{array}{c}17\\14\\12\end{array}\right\}\\2\end{array} \rightarrow \begin{array}{c}18\\14\\11\\2\end{array}$$

If $b_i - 6 \geq b_{i+1}$ then we interchange $b_{i+1}$ and $b_{i+2}$ and then attend to DC.

**eg.**
$$\begin{matrix} 22 \\ 13 \\ 10 \\ 8 \\ 2 \end{matrix} \rightarrow \begin{matrix} 22 \\ 13 \\ 18 \\ 2 \end{matrix} \rightarrow \begin{matrix} 22 \\ 18 \\ 13 \\ 2 \end{matrix}$$

*Note* 2. In this step if $(b_{i+1}, b_i - 6)$ turns out to be either a 7–5 or 4-2 pair say $(x, y)$ then we replace it by $(x + 1, y - 1)$.

**eg.** i)
$$\begin{matrix} 17 \\ 13 \\ 7 \\ 5 \\ 2 \end{matrix} \rightarrow \begin{matrix} 17 \\ 13 \\ 12 \\ 2 \end{matrix} \rightarrow \begin{matrix} 18 \\ 13 \\ 11 \\ 2 \end{matrix} \rightarrow \begin{matrix} 18 \\ 14 \\ 10 \\ 2 \end{matrix}$$
ii)
$$\begin{matrix} 23 \\ 19 \\ 13 \\ 11 \\ 2 \end{matrix} \rightarrow \begin{matrix} 23 \\ 19 \\ 24 \\ 2 \end{matrix} \rightarrow \begin{matrix} 30 \\ 19 \\ 17 \\ 2 \end{matrix} \rightarrow \begin{matrix} 30 \\ 20 \\ 16 \\ 2 \end{matrix}$$

*Case* 3. Let $b_i \not\equiv 0 \pmod 6$ and $b_{i+1}, b_{i+2} \equiv 0 \pmod 6$. Suppose $b_{i+2} + 6 \geq b_{i+1}$. Then we replace (3) by $\begin{pmatrix} b_{i+2} + 6 \\ b_{i+1} \\ b_i - 6 \end{pmatrix}$

**eg.** i)
$$\begin{matrix} 14 \\ 11 \\ 10 \\ 8 \\ 7 \end{matrix} \rightarrow \begin{matrix} 14 \\ 18 \\ 18 \end{matrix} \rightarrow \begin{matrix} 24 \\ 18 \\ 8 \end{matrix}$$
ii)
$$\begin{matrix} 16 \\ 10 \\ 8 \\ 7 \\ 5 \end{matrix} \rightarrow \begin{matrix} 16 \\ 18 \\ 12 \end{matrix} \rightarrow \begin{matrix} 18 \\ 18 \\ 10 \end{matrix}$$

If $b_{i+2} + 6 < b_{i+1}$ then we interchange $b_i$ and $b_{i+1}$ and then attend to DC.

**eg.**
$$\begin{matrix} 22 \\ 16 \\ 14 \\ 10 \\ 8 \end{matrix} \rightarrow \begin{matrix} 22 \\ 30 \\ 18 \end{matrix} \rightarrow \begin{matrix} 30 \\ 22 \\ 18 \end{matrix}$$

Suppose we come across a quadruple

$$\begin{pmatrix} b_i \\ b_{i+1} \\ b_{i+2} \\ b_{i+3} \end{pmatrix} \tag{4}$$

where $b_i, b_{i+1} \not\equiv 0 \pmod 6$, $b_{i+2}, b_{i+3} \equiv 0 \pmod 6$, $b_i - b_{i+2} < 6$ and $b_{i+1} - b_{i+3} < 6$. (4) consists of two triplets, (3) and

$$\begin{pmatrix} b_{i+1} \\ b_{i+2} \\ b_{i+3} \end{pmatrix}. \tag{5}$$

For (3) case 2 is applicable where as for (5) case 3 is applicable. Now according to the rule explained already, we have to attend to DC first for the triple (3) and then for (5). However we deviate from this rule when $b_{i+2} - b_{i+3} = 0$ or 6. In this case we attend to DC first for the triple (5) as explained in case 3.

**eg.**
$$\begin{array}{r} 19 \\ 14 \\ 18 \\ 12 \end{array} \rightarrow \left.\begin{array}{r} 19 \\ 18 \\ 18 \end{array}\right\} \begin{array}{l} \\ \\ \\ \end{array} \!\!\!\!\!\!\!\! \begin{array}{r} \\ \\ \\ 8 \end{array} \rightarrow \begin{array}{r} 24 \\ 18 \\ 13 \\ 8 \end{array}$$

(19, 14, 18, 12) comes from (19, 14, 10, 8, 7, 5) after adding the pairs.

Apply the above step $AB_4$ till DC is satisfied for all the parts of $\pi$.

**Step $AB_5$.** Finally rearrange the parts in decreasing order. It is easy to see that DC is satisfied in the rearranged partition. The resulting partition will be a partition of the type enumerated by $B^0_{5,3,3}(n)$.

We illustrate our procedure by an example.

Let $\pi = 25 + 23 + 22 + 20 + 19 + 16 + 13 + 11 + 10 + 8 + 7 + 5 + 4 + 2$ be a partition enumerated by $A_{5,3,3}(n)$.

$$\begin{array}{r} 25 \\ \left.\begin{array}{r} 23 \\ 22 \\ 20 \\ 19 \end{array}\right] 1 \\ 16 \\ 13 \\ 11 \\ 10 \\ 8 \\ 7 \\ 5 \\ 4 \\ 2 \end{array} \rightarrow \begin{array}{r} 25 \\ 42 \\ 42 \\ 16 \\ \left.\begin{array}{r} 13 \\ 11 \end{array}\right] 2 \\ \left.\begin{array}{r} 10 \\ 8 \end{array}\right] 3 \\ \left.\begin{array}{r} 7 \\ 5 \end{array}\right] 4 \\ \left.\begin{array}{r} 4 \\ 2 \end{array}\right] 5 \end{array} \rightarrow \begin{array}{r} \left.\begin{array}{r} 25 \\ 42 \\ 42 \end{array}\right] 6 \\ 16 \\ 24 \\ 18 \\ 12 \\ 6 \end{array} \rightarrow \begin{array}{r} 48 \\ 42 \\ 19 \\ \left.\begin{array}{r} 16 \\ 24 \\ 18 \end{array}\right] 7 \\ 12 \\ 6 \end{array} \rightarrow \begin{array}{r} 48 \\ 42 \\ \left.\begin{array}{r} 19 \\ 24 \\ 24 \end{array}\right] 8 \\ 10 \\ 12 \\ 6 \end{array} \rightarrow \begin{array}{r} 48 \\ 42 \\ 30 \\ 24 \\ \left.\begin{array}{r} 13 \\ 10 \\ 12 \end{array}\right] 9 \\ 6 \end{array} \rightarrow$$

$$\begin{array}{r} 48 \\ 42 \\ 30 \\ 24 \\ 18 \\ \left.\begin{array}{r} 10 \\ 7 \\ 6 \end{array}\right] 10 \end{array} \rightarrow \begin{array}{r} 48 \\ 42 \\ 30 \\ 24 \\ 18 \\ 12 \\ 7 \\ 4 \end{array}$$

Here 1–10 are respectively the following steps:

| | | | | | | | |
|---|---|---|---|---|---|---|---|
| 1) | $AB_2$ | 2) | $AB_3$ | 3) | $AB_3$ | 4) | $AB_3$ |
| 5) | $AB_3$ | 6) | $AB_4$, case 3 | 7) | $AB_4$, case 3 | 8) | $AB_4$, case 3 |
| 9) | $AB_4$, case 2 | 10) | $AB_4$, case 2. | | | | |

We now give the reverse mapping from $B^0_{5,3,3}(n)$ to $A_{5,3,3}(n)$. Let $\psi$ be a partition enumerated by $B^0_{5,3,3}(n)$. If no part of $\psi$ is a multiple of 6 then clearly it is a partition enumerated by $A_{5,3,3}(n)$, as the parts of $\psi \not\equiv 0 \pmod 6$ are all distinct. We shall explain the required map from $B^0$ to $A$ in the following steps.

**Step $BA_1$.** List the parts of $\psi$ in a column in decreasing order.

*Note* 3. In the following steps $BA_2$ and $BA_3$,

(a) each multiple of 6, say $12j$ or $12j+6$ is split into a pair $(\alpha, \beta)$ as follows:

$$\left.\begin{array}{ll} 12j & \to (6j+1, 6j-1),\ j \geq 1 \\ 12j+6 & \to (6j+4, 6j+2),\ j \geq 0 \end{array}\right\} = (\alpha, \beta),\ (\text{say}) \tag{6}$$

**eg.** i) $12 \to \begin{array}{c} 7 \\ 5 \end{array}$ ii) $18 \to \begin{array}{c} 10 \\ 8 \end{array}$

(b) after splitting each multiple of 6, (say $x$) into a pair $(\alpha, \beta)$ rearrange the parts below $x$ in decreasing order.

**eg.** i) $\left.\begin{array}{c} 123 \\ 63 \\ 56 \\ 51 \end{array}\right\} \to \begin{array}{c} 64 \\ 62 \\ 63 \\ 56 \\ 51 \end{array} \to \begin{array}{c} 64 \\ 63 \\ 62 \\ 56 \\ 51 \end{array}$

**Step $BA_2$.** Starting from the **bottom**, look for a multiple of 6, say $12j$ or $12j+6$. If $\alpha$ or $\beta$ is not a part lying below it, then we split the multiple of 6 as in (6)

**eg.** i) $\begin{array}{c} 17 \\ 12 \\ 2 \end{array} \to \begin{array}{c} 17 \\ 7 \\ 5 \\ 2 \end{array}$ ii) $\begin{array}{c} 18 \\ 4 \\ 1 \end{array} \to \begin{array}{c} 10 \\ 8 \\ 4 \\ 1 \end{array}$

If $\alpha$ or $\beta$ is a part lying below the multiple of 6 then we split it in the order given below (multiples of 6 are omitted).

$$\left.\begin{array}{lll} 12j & \to (6j+2, 6j-2), \ldots (12j-2, 2), (12j-1, 1), & j \geq 1 \\ 12j+6 & \to (6j+5, 6j+1) \ldots (12j+4, 2), (12j+5, 1), & j \geq 0 \end{array}\right\} \tag{7}$$

**eg.** i) $\begin{array}{c} 12 \\ 5 \\ 2 \end{array} \to \begin{array}{c} 8 \\ 4 \\ 5 \\ 2 \end{array} \to \begin{array}{c} 8 \\ 5 \\ 4 \\ 2 \end{array}$ ii) $\begin{array}{c} 18 \\ 10 \\ 5 \end{array} \to \begin{array}{c} 11 \\ 7 \\ 10 \\ 5 \end{array} \to \begin{array}{c} 11 \\ 10 \\ 7 \\ 5 \end{array}$

**Step $BA_3$.** From the bottom look for a triple (3) in which either $b_i$ or both $b_i$ and $b_{i+1}$ are multiples of 6. Then we have the following possibilities:

*Case* 1. Suppose $b_i \equiv 0 (\text{mod } 6)$ and $b_{i+1}, b_{i+2} \not\equiv 0 (\text{mod } 6)$.Let $(\alpha, \beta)$ be the corresponding pair of $b_i$. Then we replace (3) by

$$\begin{pmatrix} \alpha \\ \beta \\ b_{i+1} \\ b_{i+2} \end{pmatrix} \quad \text{if} \quad 1 \leq b_{i+2} < \beta - 1$$

**eg.** i) $\begin{matrix} 18 \\ 5 \\ 3 \end{matrix} \rightarrow \begin{matrix} 10 \\ 8 \\ 5 \\ 3 \end{matrix}$ ii) $\begin{matrix} 24 \\ 8 \\ 4 \end{matrix} \rightarrow \begin{matrix} 13 \\ 11 \\ 8 \\ 4 \end{matrix}$

If $\beta - 1 \leq b_{i+2} \leq b_i - 7$ and $b_{i+2} + 6 > b_{i+1}$ then replace (3) by

$$\begin{pmatrix} b_{i+2} + 6 \\ b_{i+1} \\ b_i - 6 \end{pmatrix}. \tag{8}$$

Let $(\alpha', \beta')$ be the corresponding pair of $b_i - 6$. If $\alpha'$ or $\beta'$ is not a part below $b_i - 6$ then it is split as in (6). If $\alpha'$ or $\beta'$ is a part below $b_i - 6$ then we split it in the order stated in (7).

**eg.** i) $\begin{matrix} 24 \\ 14 \\ 11 \\ 5 \end{matrix} \rightarrow \begin{matrix} 17 \\ 14 \\ 18 \\ 5 \end{matrix} \rightarrow \begin{matrix} 17 \\ 14 \\ 10 \\ 8 \\ 5 \end{matrix}$ ii) $\begin{matrix} 24 \\ 14 \\ 11 \\ 8 \end{matrix} \rightarrow \begin{matrix} 17 \\ 14 \\ 18 \\ 8 \end{matrix} \rightarrow \begin{matrix} 17 \\ 14 \\ 11 \\ 8 \\ 7 \end{matrix}$

If $\beta - 1 \leq b_{i+2} \leq b_i - 7$ and $b_{i+2} + 6 \leq b_{i+1}$ then we interchange $b_i$ and $b_{i+1}$ and again apply the same procedure.

**eg.** $\begin{matrix} 18 \\ 15 \\ 8 \end{matrix} \rightarrow \begin{matrix} 15 \\ 18 \\ 8 \end{matrix} \rightarrow \begin{matrix} 15 \\ 11 \\ 8 \\ 7 \end{matrix}$

*Note* 4. In (8) if $(b_{i+2} + 6, b_{i+1})$ turns out to be a 7–5 or 4–2 pair then we replace it by its next pair in the order stated in (7).

**eg.** i) $\begin{matrix} 12 \\ 8 \\ 4 \end{matrix} \rightarrow \begin{matrix} 10 \\ 8 \\ 6 \end{matrix} \rightarrow \begin{matrix} 11 \\ 7 \\ 6 \end{matrix} \rightarrow \begin{matrix} 11 \\ 7 \\ 4 \\ 2 \end{matrix}$ ii) $\begin{matrix} 24 \\ 17 \\ 13 \end{matrix} \rightarrow \begin{matrix} 19 \\ 17 \\ 18 \end{matrix} \rightarrow \begin{matrix} 20 \\ 16 \\ 18 \end{matrix} \rightarrow \begin{matrix} 20 \\ 16 \\ 10 \\ 8 \end{matrix}$

*Case* 2. Suppose there exists a triple (3) where $b_i, b_{i+1} \equiv 0 \pmod 6$ and $b_{i+2} \not\equiv 0 \pmod 6$. Let $(\alpha_1, \beta_1)$ and $(\alpha_2, \beta_2)$ be the corresponding pairs of $b_i$ and $b_{i+1}$ respectively.

If $1 \le b_{i+2} < \beta_1 - 1$ then (3) is first replaced by

$$\begin{pmatrix} b_i \\ \alpha_2 \\ \beta_2 \\ b_{i+2} \end{pmatrix}.$$

If $\alpha_2$ or $\beta_2$ is one of the parts in $b_{i+2}, \dots, b_s$ then we replace $(\alpha_2, \beta_2)$ by its next pair in the order stated in (7).

Finally we replace the above quadruple by

$$\begin{pmatrix} \alpha_1 \\ \beta_1 \\ \alpha_2 \\ \beta_2 \\ b_{i+2} \end{pmatrix}.$$

If $\alpha_1$ or $\beta_1$ is one of the parts in $\alpha_2, \beta_2, b_{i+2}, \dots, b_s$ then we replace $(\alpha_1, \beta_1)$ by its next pair in the order stated in (7).

**eg.** i) $$\begin{matrix} 15 \\ 10 \\ 2 \\ 1 \end{matrix} \to \begin{matrix} 18 \\ 12 \\ 2 \\ 1 \end{matrix} \to \begin{matrix} 18 \\ 7 \\ 5 \\ 2 \\ 1 \end{matrix} \to \begin{matrix} 10 \\ 8 \\ 7 \\ 5 \\ 2 \\ 1 \end{matrix}$$ ii) $$\begin{matrix} 30 \\ 24 \\ 11 \end{matrix} \to \begin{matrix} 30 \\ 14 \\ 10 \\ 11 \end{matrix} \to \begin{matrix} 30 \\ 14 \\ 11 \\ 10 \end{matrix} \to \begin{matrix} 17 \\ 13 \\ 14 \\ 11 \\ 10 \end{matrix} \to \begin{matrix} 17 \\ 14 \\ 13 \\ 11 \\ 10 \end{matrix}$$

If $\beta_1 - 1 \le b_{i+2} \le b_i - 7$ and $b_i - 6 \le b_{i+1}$ then we replace (3) by $\begin{pmatrix} b_{i+2} + 6 \\ b_{i+1} \\ b_i - 6 \end{pmatrix}$.

**eg.** $$\begin{matrix} 18 \\ 12 \\ 8 \end{matrix} \to \begin{matrix} 14 \\ 12 \\ 12 \end{matrix}$$

If $\beta_1 - 1 \le b_{i+2} \le b_i - 7$ and $b_i - 6 > b_{i+1}$ then we interchange $b_{i+1}$ and $b_{i+2}$ and continue the process.

**eg.** i) $$\begin{matrix} 24 \\ 12 \\ 11 \end{matrix} \to \begin{matrix} 24 \\ 11 \\ 12 \end{matrix} \to \begin{matrix} 24 \\ 11 \\ 7 \\ 5 \end{matrix} \to \begin{matrix} 14 \\ 10 \\ 11 \\ 7 \\ 5 \end{matrix} \to \begin{matrix} 14 \\ 11 \\ 10 \\ 7 \\ 5 \end{matrix}$$

**Particular case of case 2** ($PC_2$).

If $b_i, b_{i+1} \equiv 0 (\mathrm{mod}\ 6)$ and $b_i = b_{i+1}$ then we replace (3) by

$$\begin{pmatrix} \alpha + 1 \\ \alpha \\ \beta \\ \beta - 1 \\ b_{i+2} \end{pmatrix} \quad \text{if } 1 \le b_{i+2} < \beta - 1.\ (\alpha, \beta) \text{ is the pair of } b_i (= b_{i+1}).$$

**eg.** i) $\begin{matrix} 18 \\ 18 \\ 4 \end{matrix} \rightarrow \begin{matrix} 11 \\ 10 \\ 8 \\ 7 \\ 4 \end{matrix}$ ii) $\begin{matrix} 18 \\ 12 \\ 8 \end{matrix} \rightarrow \begin{matrix} 14 \\ 12 \\ 12 \end{matrix} \rightarrow \begin{matrix} 14 \\ 8 \\ 7 \\ 5 \\ 4 \\ 2 \end{matrix}$

If $\beta - 1 \le b_{i+2} \le b_i - 7$ then the procedure explained in case 2 is applied.

**eg.** $\begin{matrix} 24 \\ 24 \\ 11 \end{matrix} \rightarrow \begin{matrix} 17 \\ 24 \\ 18 \end{matrix} \rightarrow \begin{matrix} 17 \\ 24 \\ 10 \\ 8 \end{matrix} \rightarrow \begin{matrix} 17 \\ 13 \\ 11 \\ 10 \\ 8 \end{matrix}$

Suppose there exists a quadruple

$$\begin{pmatrix} b_i \\ b_{i+1} \\ b_{i+2} \\ b_{i+3} \end{pmatrix} \tag{9}$$

where $b_i, b_{i+1} \equiv 0 (\mathrm{mod}\ 6)$ and $b_{i+2}, b_{i+3} \not\equiv 0 (\mathrm{mod}\ 6)$. (9) consists of two triplets, (3) and

$$\begin{pmatrix} b_{i+1} \\ b_{i+2} \\ b_{i+3} \end{pmatrix}. \tag{10}$$

For (3) case 2 is applicable where as for (10) case 1 is applicable. According to the procedure explained above, we have to attend first to the triple (10) and then to (3). But we deviate from this rule in the following two cases:

i) $b_i - b_{i+1} = 0$
ii) $b_i - b_{i+1} = 6$ and $\beta - 1 \le b_{i+2} \le b_i - 7$ where $(\alpha, \beta)$ is the corresponding pair of $b_i$.

In these two cases we attend first to the triple (3) as explained in case 2 and then to (10).

**eg.** i)

$$\begin{array}{c}\left.\begin{array}{c}30\\24\\16\end{array}\right\}\\13\end{array}\rightarrow\begin{array}{c}22\\\left.\begin{array}{c}24\\24\\13\end{array}\right\}\end{array}\rightarrow\begin{array}{c}22\\19\\24\\18\end{array}\rightarrow\begin{array}{c}22\\19\\24\\10\\8\end{array}\rightarrow\begin{array}{c}22\\19\\13\\11\\10\\8\end{array}$$

ii)

$$\begin{array}{c}30\\\left.\begin{array}{c}24\\11\\10\end{array}\right\}\end{array}\rightarrow\begin{array}{c}30\\16\\11\\18\end{array}\rightarrow\begin{array}{c}\left.\begin{array}{c}30\\16\\11\end{array}\right\}\\10\\8\end{array}\rightarrow\begin{array}{c}17\\13\\16\\11\\10\\8\end{array}\rightarrow\begin{array}{c}17\\16\\13\\11\\10\\8\end{array}$$

From the definition of $A_{5,3,3}(n)$ it is clear that no part is $\equiv 0 (\text{mod } 3)$. After adding the pairs we get parts $\equiv 0(\text{mod } 6)$ which is true even in the case of adding the parts of a quadruple of the type (1) and (2). Hence, it is clear that in the resulting partitions enumerated by $B^0_{5,3,3}(n)$, no part is $\equiv 3(\text{mod } 6)$. Also from the map $A \longrightarrow B^0$ which we have explained, it is clear from step $AB_3$ that there will be no 7–5 or 4–2 pairs and from step $AB_4$ it follows that DC is taken care of. So we have only to check whether the resulting partitions (in order to be enumerated by $B^0$) satisfy the condition:

$$S : f_{6j-1} + f_{6j} + f_{6j+6} + f_{6j+7} \leq 3, \quad j \geq 1.$$

We now prove this.

*Proof.* Assume the contrary that $S$ is not satisfied. Then there exists an integer $j \geq 1$ for which we can find a quadruple

$$(6j + 7, 6j + 6, 6j, 6j - 1) \tag{11}$$

in a partition $\pi$ enumerated by $B^0_{5,3,3}(n)$. Let $\pi^1$ and $\pi^2$ denote respectively the partitions in $\pi$ with parts $> 6j + 7$ and $< 6j - 1$ which do not contain quadruples of the type (11). If we arrange the parts of $\pi$ in decreasing order then we get

$$\begin{array}{c}\pi^1\\6j+7\\6j+6\\6j\\6j-1\\\pi^2\end{array}.$$

According to the procedure explained for the reverse map this partition goes to the following partition. The steps are as follows:

$$
\psi^1 : \begin{array}{c} \pi^1 \\ 6j+7 \\ \left.\begin{array}{c} 6j+6 \\ 6j \\ 6j-1 \end{array}\right\} BA_3,\ \text{case 2} \\ \pi^2 \end{array}
\longrightarrow
\psi^2 : \begin{array}{c} \pi^{1*} \\ 6j+7 \\ 6j+5 \\ \left.\begin{array}{c} 6j \\ 6j \end{array}\right\} BA_3,\ PC_2 \\ \pi^{2*} \end{array}
\longrightarrow
\psi^3 : \begin{array}{c} \pi^{1*} \\ 6j+7 \\ 6j+5 \\ \alpha+1 \\ \alpha \\ \beta \\ \beta-1 \\ \pi^{2*} \end{array}
$$

where $(\alpha, \beta)$ is the pair of $5j$ and $\pi^{1*}$ and $\pi^{2*}$ are the partitions obtained from $\pi^1$ and $\pi^2$ respectively after applying the procedure. We observe that $\psi^3$ is enumerated by $A_{5,3,3}(n)$.

But if we apply the procedure explained for the map from $A$ to $B^0$, then $\psi^3$ is mapped on to a totally different partition. We now give the details.

$$
\psi^3 : \begin{array}{c} \pi^{1*} \\ 6j+7 \\ 6j+5 \\ \alpha+1 \\ \alpha \\ \beta \\ \beta-1 \\ \pi^{2*} \end{array}
\longrightarrow
\begin{array}{c} \pi^1 \\ 6j+7 \\ 6j+5 \\ 6j \\ 6j \\ \pi^2 \end{array}
\longrightarrow
\begin{array}{c} \pi^1 \\ 12j+12 \\ 6j \\ 6j \\ \pi^2 \end{array}
$$

Similar proof can be given if either $\pi^1$ or $\pi^2$ or both contain quadruples of the type (11).

We now illustrate the reverse map by considering the same partition $\psi = 48 + 42 + 30 + 24 + 18 + 12 + 7 + 4$ obtained from $\pi$.

$$
\begin{array}{c} 48 \\ 42 \\ 30 \\ 24 \\ \left.\begin{array}{c} 18 \\ 12 \\ 7 \end{array}\right] 1 \\ 4 \end{array}
\rightarrow
\begin{array}{c} 48 \\ 42 \\ 30 \\ 24 \\ 13 \\ \left.\begin{array}{c} 12 \\ 12 \\ 4 \end{array}\right] 2 \end{array}
\rightarrow
\begin{array}{c} 48 \\ 42 \\ 30 \\ 24 \\ 13 \\ 10 \\ 12 \\ 6 \end{array}
\rightarrow
\begin{array}{c} 48 \\ 42 \\ 30 \\ 24 \\ 13 \\ 10 \\ \left.\begin{array}{c} 12 \\ 4 \\ 2 \end{array}\right] 3 \end{array}
\rightarrow
\begin{array}{c} 48 \\ 42 \\ \left.\begin{array}{c} 30 \\ 24 \\ 13 \end{array}\right] 4 \\ 10 \\ 7 \\ 5 \\ 4 \\ 2 \end{array}
\rightarrow
\begin{array}{c} 48 \\ 42 \\ 19 \\ \left.\begin{array}{c} 24 \\ 24 \\ 10 \end{array}\right] 5 \\ 7 \\ 5 \\ 4 \\ 2 \end{array}
\rightarrow
$$

$$\begin{array}{c} 48\\ 42\\ 19\\ 16\\ 24\\ \left.\begin{array}{c}18\\ 7\\ 5\end{array}\right]6\\ 4\\ 2 \end{array} \rightarrow \begin{array}{c} 48\\ 42\\ 19\\ 16\\ \left.\begin{array}{c}24\\ 10\\ 8\end{array}\right]7\\ 7\\ 5\\ 4\\ 2 \end{array} \rightarrow \begin{array}{c} 48\\ \left.\begin{array}{c}42\\ 19\\ 16\end{array}\right]8\\ 13\\ 11\\ 10\\ 8\\ 7\\ 5\\ 4\\ 2 \end{array} \rightarrow \begin{array}{c} \left.\begin{array}{c}48\\ 22\\ 20\end{array}\right]9\\ 19\\ 16\\ 13\\ 11\\ 10\\ 8\\ 7\\ 5\\ 4\\ 2 \end{array} \rightarrow \begin{array}{c} 25\\ 23\\ 22\\ 20\\ 19\\ 16\\ 13\\ 11\\ 10\\ 8\\ 7\\ 5\\ 4\\ 2 \end{array}$$

Here 1–9 are respectively the following steps:

| | | | |
|---|---|---|---|
| 1) $BA_3$, case 2 | 2) $BA_3$, case 2 | 3) $BA_3$, case 1 | 4) $BA_3$, case 2 |
| 5) $BA_3$, case 2 | 6) $BA_3$, case 1 | 7) $BA_3$, case 1 | 8) $BA_3$, case 1 |
| 9) $BA_3$, case 1. | | | |

We now show that the maps from $A \longrightarrow B^0$ and $B^0 \longrightarrow A$ that we have described are reversible. The following list proves this.

1) Step $AB_1$ and step $BA_1$ are same.
2) Step $AB_2$ is reversible to step $BA_3$, case 2.
3) Step $AB_3$ is reversible to cases 1,2 of step $BA_3$.
4) In step $AB_4$,
   case 1 is reversible to step $BA_2$
   case 2 is reversible to step $BA_3$, case 1
   case 3 is reversible to step $BA_3$, case 2.
5) Step $AB_5$ is just rearranging the parts in decreasing order. But in the map $B^0 \longrightarrow A$ this is taken care of by Note 3(b).

## 3. Table

| $n$ | $A_{5,3,3}(n)$ | $n$ | $A_{5,3,3}(n)$ | $n$ | $A_{5,3,3}(n)$ |
|---|---|---|---|---|---|
| 1 | 0 | 21 | 20 | 41 | 187 |
| 2 | 1 | 22 | 23 | 42 | 207 |
| 3 | 1 | 23 | 26 | 43 | 228 |
| 4 | 1 | 24 | 30 | 44 | 250 |
| 5 | 2 | 25 | 34 | 45 | 274 |
| 6 | 2 | 26 | 38 | 46 | 301 |
| 7 | 3 | 27 | 42 | 47 | 331 |
| 8 | 3 | 28 | 47 | 48 | 364 |
| 9 | 3 | 29 | 53 | 49 | 399 |
| 10 | 4 | 30 | 60 | 50 | 436 |
| 11 | 5 | 31 | 67 | 51 | 476 |
| 12 | 6 | 32 | 74 | 52 | 520 |
| 13 | 7 | 33 | 82 | 53 | 569 |
| 14 | 8 | 34 | 91 | 54 | 622 |
| 15 | 9 | 35 | 102 | 55 | 679 |
| 16 | 10 | 36 | 114 | 56 | 739 |
| 17 | 12 | 37 | 126 | 57 | 804 |
| 18 | 14 | 38 | 139 | 58 | 875 |
| 19 | 16 | 39 | 153 | 59 | 953 |
| 20 | 18 | 40 | 169 | 60 | 1038 |

## References

[A1] G. E. Andrews, A Generalization of the Classical Parition Theorems, *Trans. Amer. Math. Soc.* **145** (1969) 205–221

[A2] G. E. Andrews, On the General Rogers – Ramanujan Theorem, *Mem. Amer. Math. Soc.* no. 152 (1974) 1–86

[ABO] G. E. Andrews, C. Bessenrodt and J. B. Olsson, Partition Identities and Labels for Modular Characters, *Trans. Amer. Math. Soc.*, Vol. **344** no. 2, August (1994) 597–615

[PR] Padmavathamma and M. Ruby Salestina, A combinatorial proof of a theorem of Andrews on Partitions – Communicated

[S] I. J. Schur, Zur additiven Zahlentheorie, Sitzungsber. Deutsch. Adak. Wissensch. Berlin, *Phys-Math.* **K1** (1926) 488–495

Proc. Int. Conf. – Number Theory No. 1, 2004, pp. 71–88.

# A Survey on $k$–Freeness

Francesco Pappalardi

Dipartimento di Matematica, Università Roma Tre, Largo S. L. Murialdo 1, Roma, 00146 — Italy
e-mail: pappa@mat.uniroma3.it

June 14, 2004

**Abstract.** We say that an integer $n$ is $k$–free ($k \geq 2$) if for every prime $p$ the valuation $v_p(n) < k$. If $f : \mathbb{N} \to \mathbb{Z}$, we consider the enumerating function $\mathcal{S}_f^k(x)$ defined as the number of positive integers $n \leq x$ such that $f(n)$ is $k$–free. When $f$ is the identity then $\mathcal{S}_f^k(x)$ counts the $k$–free positive integers up to $x$. We review the history of $\mathcal{S}_f^k(x)$ in the special cases when $f$ is the identity, the characteristic function of an arithmetic progression a polynomial, arithmetic. In each section we present the proof of the simplest case of the problem in question using exclusively elementary or standard techniques.

## 1. Introduction - the classical problem

We say that an integer $n \in \mathbb{N}$ is *square free* if for any prime $p \mid n$, one has $p^2 \nmid n$. If $\mu$ denotes the Möbius function, then $\mu^2$ is the characteristic function of the set of square free numbers. It is a classical statement that

$$\mathcal{S}(x) := \#\{n \leq x \mid n \text{ is square free }\} = \frac{6}{\pi^2}x + O(\sqrt{x}).$$

The proof is simple and it goes as follows: we start from the identity

$$\mu^2(n) = \sum_{d^2 \mid n} \mu(d),$$

which follows from the fact that $\mu$ is multiplicative. From this we obtain

$$\begin{aligned}\mathcal{S}(x) &= \sum_{n \leq x} \mu^2(n) = \sum_{n \leq x} \sum_{d^2 \mid n} \mu(d)\\ &= \sum_{d \leq \sqrt{x}} \mu(d) \left(\frac{x}{d^2} + O(1)\right) = x \sum_{d \leq \sqrt{x}} \frac{\mu(d)}{d^2} + O(\sqrt{x})\\ &= x \sum_{d=1}^{\infty} \frac{\mu(d)}{d^2} + O\left(\sqrt{x} + x \sum_{d > \sqrt{x}} \frac{1}{d^2}\right) = \frac{x}{\zeta(2)} + O(x^{1/2})\end{aligned}$$

where $\zeta$ denotes the Riemann zeta function.

Similarly, if $k \geq 2$, we say that an integer $n \in \mathbb{N}$ is $k$–free if for each prime $p \mid n$, one has $p^k \nmid n$. If $\mu^{(k)}$ denotes the characteristic function of $k$–free integers, then one has the identity:

$$\mu^{(k)}(n) = \sum_{d^k|n} \mu(d).$$

The same proof as above gives:

$$\mathcal{S}^k(x) := \#\{n \leq x \mid n \text{ is } k\text{–free}\} = \frac{x}{\zeta(k)} + O(\sqrt[k]{x}).$$

To improve the error term one can consider the generating zeta function of $\mu^{(k)}$. Indeed, it is quite simple to verify that for each $s \in \mathbb{C}$ with $\Re(s) > 1$,

$$\sum_{n=1}^{\infty} \frac{\mu^{(k)}(n)}{n^s} = \frac{\zeta(s)}{\zeta(ks)}.$$

Therefore the right hand side above can be extended to a meromorphic function on $\mathbb{C}$ with poles on $s = 1$ and on the complex numbers of the form $\rho/k$ where $\rho$ ranges over the zeros of $\zeta(s)$.

From the Perron integral (see the book of G. Tenenbaum: Introduction to analytic and probabilistic number theory, 1995 on page 130):

$$\frac{1}{2\pi i} \int_{\Re(s)=2} \frac{y^s}{s} ds = \begin{cases} 0 & \text{if } y < 1 \\ 1/2 & \text{if } y = 1 \\ 1 & \text{if } y > 1, \end{cases}$$

we deduce that if $x \notin \mathbb{N}$,

$$\mathcal{S}^k(x) = \frac{1}{2\pi i} \int_{\Re(s)=2} \frac{\zeta(s)}{\zeta(ks)} \frac{x^s}{s} ds.$$

Computing the residues, we obtain:

$$\mathcal{S}^k(x) = \frac{x}{\zeta(k)} + \sum_{\substack{\rho \\ \zeta(\rho)=0}} a_{\rho,k} x^{\rho/k}.$$

where $a_{\rho,k}$ can be computed in terms of the residue of $1/\zeta(s)$ at $s = \rho$.

If we assume the Riemann Hypothesis and we write $\rho = 1/2 + i\gamma$, then the above identity becomes

$$\mathcal{S}^k(x) = \frac{x}{\zeta(k)} + O\left(x^{1/2k} \sum_{\gamma} a_{\rho,k} x^{i\gamma/k}\right).$$

This argument suggests that the right error term in an asymptotic formula for the number of $k$–free integers corresponds to the largest real part of the zeroes of the Riemann $\zeta$–function. Therefore the following conjecture should hold:

$$\mathcal{S}^k(x) = \frac{x}{\zeta(k)} + O(x^{1/2k+\epsilon}).$$

The above statement has a clear connection with the Riemann Hypothesis and therefore it is quite unlikely that it will be proven in the near future.

Using oscillation theorems, it is also possible to prove that

$$R_k(x) := \mathcal{S}^k(x) - \frac{x}{\zeta(k)} = \Omega(\sqrt[2k]{x}).$$

This was shown in 1968 by Vaidya [38].

In this direction Balasubramanian and Ramachandra in 1988 [3,4] gave an effective proof that

$$R_2(x) = \Omega_{\pm}(\sqrt{x}).$$

The upper bound for $R_k(x)$ has attracted the work of many authors. Not much can be said unconditionally except that, using the classical zero–free region estimates due to Vinogradov and Korobov through exponential sums, Walfisz [40] showed that for a suitable constant $c$, uniformly on $k$,

$$R_k(x) \ll \sqrt[k]{x}\exp\{-ck^{-8/5}\log^{3/5} x \log\log^{1/5} x\}.$$

We assume the Riemann Hypothesis for the rest of this section and list the achievements on the problem of estimating $R_k(x)$:

- In 1911 Axer [1] showed that for every $\epsilon > 0$

$$R_k(x) \ll x^{2/(2k+1)+\epsilon}.$$

- In 1981 Montgomery and Vaughan [27] showed that for every $\epsilon > 0$

$$R_k(x) \ll x^{1/(k+1)+\epsilon}.$$

  In the same paper, they improved the above in the case $k = 2$ showing that for every $\epsilon > 0$

$$R_2(x) \ll x^{9/28+\epsilon}.$$

- In the same year the exponent above was improved by Graham [11] showing for every $\epsilon > 0$

$$R_2(x) \ll x^{8/25+\epsilon}.$$

- In 1985 Baker and Pintz [2] showed that for every $\epsilon > 0$

$$R_2(x) \ll x^{7/22+\epsilon}.$$

The same result was also obtained by Jia in 1987 [22].

- In 1984 Yao [39] showed that for every $\epsilon > 0$

$$R_k(x) \ll x^{9/(9k+7)+\epsilon}.$$

- In 1988 Li [24] showed that for every $\epsilon > 0$

$$R_k(x) \ll x^{\max\{2/(2k+3), 9/(10k+8)+\epsilon\}}.$$

- In 1989 Graham and Pintz [13] showed that for every $\epsilon > 0$

$$R_k(x) \ll x^{D(k)+\epsilon},$$

where

$$D(k) = \begin{cases} 7/(8k+6), & \text{if } 2 \le k \le 5; \\ \frac{67}{514} & \text{if } k = 6; \\ 11(k-4)/(12k^2-37k-41) & \text{if } 7 \le k \le 12; \\ 23(k-1)/(24k^2+13k-37) & \text{if } 13 \le k \le 20. \end{cases}$$

For $k \ge 21$ the definition of $D(k)$ is more complicated, but it may be shown that $D(k) \sim 2\log 2/(k + \log k)$ for $k \to \infty$. These estimates could be improved for large $k$ by appealing to Vinogradov's method. The authors obtain $R_k(x) \ll x^{E(k)}$, where $E(k) = 1/(k + ck^{1/3})$ for some constant $c$.

- The most recent contribution to our knowledge is due to Jia [23] in 1993. He showed that for every $\epsilon > 0$

$$R_2(x) \ll x^{17/54+\epsilon}.$$

## 2. $k$–free numbers in arithmetic progressions

Let $q \in \mathbb{N}$ and let $a \in \mathbb{Z}/q\mathbb{Z}$ be such that $\gcd(a, q)$ is $k$–free, then we set

$$S_k(x; a, q) := \#\{n \le x \mid n \text{ is } k\text{–free and } n \equiv a \bmod q\}.$$

The first result about the distribution of $k$–free numbers in arithmetic progressions that we would like to mention is due to Prachar in 1958 [35]. He proved that if $(a, q) = 1$ and if

$$\delta_{k,q} := \frac{1}{q\zeta(k)} \prod_{l|q} \left(1 - \frac{1}{l^k}\right),$$

then

$$\mathcal{S}_k(x; a, q) = \delta_{k,q} x + O(\sqrt[k]{x} q^{-1/k^2} + q^{1/k} k^{\omega(q)}),$$

where $\omega(q)$ denotes the number of prime divisors of $q$. Prachar's proof goes as follows: using the formula

$$\mu^{(k)}(n) = \sum_{d^k|n} \mu(d),$$

and interchanging the order of summation, we write

$$\mathcal{S}_k(x; a, q) = \sum_{d \le \sqrt[k]{x}} \mu(d) \#\{m \le x/d^k \mid d^k m \equiv a \bmod q\}. \tag{1}$$

Observing that since $\gcd(a, q) = 1$, the sum above is only supported at values of $d$ such that $\gcd(d, q) = 1$. Furthermore

$$\#\{m \le x/d^k \mid d^k m \equiv a \bmod q\} = \frac{x}{qd^k} + O(1).$$

Let us split the sum in (1) in two sums; the first with the values $d \le \sqrt[k]{x}/q^{1/k^2}$ and the second with the values $\sqrt[k]{x}/q^{1/k^2} < d \le \sqrt[k]{x}$.

For the first sum note that

$$\begin{aligned}
&\sum_{\substack{d \le \sqrt[k]{x}/q^{1/k^2} \\ \gcd(d,q)=1}} \mu(d) \#\{m \le x/d^k \mid d^k m \equiv a \bmod q\} \\
&= \sum_{\substack{d \le \sqrt[k]{x}/q^{1/k^2} \\ \gcd(d,q)=1}} \mu(d) \left(\frac{x}{qd^k} + O(1)\right) \\
&= \frac{x}{q} \sum_{\substack{d=1 \\ \gcd(d,q)=1}}^{\infty} \frac{\mu(d)}{d^k} + O\left(\frac{x}{q} \sum_{d > \sqrt[k]{x}/q^{1/k^2}} \frac{1}{d^k} + \frac{\sqrt[k]{x}}{q^{1/k^2}}\right) \\
&= \frac{x}{q\zeta(k)} \prod_{l|q} \left(1 - \frac{1}{l^k}\right) + O\left(\frac{\sqrt[k]{x}}{q^{1/k^2}}\right) \\
&= \delta_{k,q} x + O\left(\frac{\sqrt[k]{x}}{q^{1/k^2}}\right)
\end{aligned}$$

since $1 - (k-1)/k^2 > 1/k^2$.

For the second sum note that if $\sqrt[k]{x}/q^{1/k^2} < d \leq \sqrt[k]{x}$ and $nd^k \leq x$, then necessarily $n \leq q^{1/k}$. Therefore

$$\sum_{\substack{\frac{\sqrt[k]{x}}{q^{1/k^2}} < d \leq \sqrt[k]{x} \\ \gcd(d,q)=1}} \sum_{\substack{n \leq x/d^k \\ nd^k \equiv a \bmod q}} \mu(d) \ll \sum_{n \leq q^{1/k}} \#\{d \leq \left(\frac{x}{n}\right)^{1/k} \mid d^k n \equiv a \bmod q\}. \qquad (2)$$

Next note that for a fixed $n$, by the Chinese remainder Theorem the congruence $d^k n \equiv a \bmod q$ has at most $2k^{\omega(q)}$ solutions $d \in \mathbb{Z}/q\mathbb{Z}$. Hence

$$\#\{d \leq (x/n)^{1/k} \mid d^k n \equiv a \bmod q\} \leq 2k^{\omega(q)} \left( \left(\frac{x}{n}\right)^{1/k} \frac{1}{q} + O(1) \right).$$

Finally (2) is

$$O\left(k^{\omega(q)} \left(q^{1/k} + \frac{x^{1/k}}{q^{1+1/(k^2-k)}}\right)\right).$$

The above together with the estimate for the first sum complete the proof of Prachar's statement.

In the more general situation when $\gcd(a, q) > 1$, in order to exist a $k$-free number $n \equiv a \bmod q$, $\gcd(a, q)$ must be $k$–free. In this case the density of such $n$ depends upon $q$.

An asymptotic formula of the type

$$\mathcal{S}_k(x; a, q) \sim \delta_{k,a,q} x$$

in this general case appeared in the book of Landau (Handbuch der Lehre von der Verteilung der Primzahlen, 1953) for $k = 2$ and Ostmann (Additive Zahlentheorie, 1956) for $k > 2$. The value of the density is given by

$$\delta_{k,a,q} = \frac{\varphi(q)}{(a,q)\varphi(q/(a,q))} \frac{1}{\zeta(k) q} \prod_{l \mid q} \left(1 - \frac{1}{l^k}\right)^{-1}.$$

An error term for the general case has been worked out by Cohen and Robinson in 1963 [9].

By comparing the main term in Prachar's Theorem with the error term, we deduce that when $k$ is fixed, the asymptotic formula of Prachar holds uniformly for $q \leq x^{2/3-\epsilon}$ since $k^{\omega(q)} \ll q^{\epsilon}$. In 1975 Hooley [15] proves that if $\gcd(a, q) = 1$, then

$$\mathcal{S}_2(x; a, q) = \delta_{2,q} x + O(\sqrt{x} q^{-1/2} + q^{1/2+\epsilon})$$

improving Prachar's result in the range $x^{1/3} < q < x^{2/3-\epsilon}$. In the same paper Hooley proves that for every $5/8 \leq \alpha < 3/4$ there exists a constant $\eta = \eta(\alpha)$ such that

$$\mathcal{S}_2(x; a, q) = \delta_{2,q} x + O\left(\left(\frac{x}{q}\right)^{1-\eta}\right)$$

for a positive proportion of $q \in (Q, 2Q)$ provided that $x^{5/8} < Q \leq x^{\alpha}$ and $\gcd(a, q) = 1$.

A further contribution to this problem in the case $\gcd(a, q) = 1$ was proposed by McCurley in 1982 [26] when he proved that there exist absolute computable constants $c_1$ and $c_2$ such that if $q$ is such that none of the $L$-functions associated to real characters modulo $q$ have a Siegel zero, then

$$\mathcal{S}_k(x; a, q) = \delta_{k,q}x + O\left(\frac{(xq)^{1/k}}{\exp(c_1\sqrt{\log x/k^3})}\right) \quad \text{if } x \geq \exp(c_2 k \log^2 q).$$

Furthermore let $q$ and $k$ be such that $L(s, \chi^k)$ has no Siegel zeros for all $\chi$ modulo $q$. Then there exist constants $c_3$ and $c_4$ such that

$$\mathcal{S}_k(x; a, q) = \delta_{k,q}x + O\left(\frac{x^{1/k}}{\exp(c_3\sqrt{\log x/k^3})}\right) \quad \begin{array}{l}\text{if } x \geq \exp(c_4 k \log^2 q)\\ \text{and } (a, q) = 1.\end{array}$$

Average type results for $k$–free numbers in arithmetic progressions have also been considered. Orr in 1971 [31] proved the following two statements.

For any constant $A$

$$\sum_{q \leq x^{2/3}/\log^{A+1} x} \max_{\substack{a \in \mathbb{Z}/q\mathbb{Z} \\ \gcd(a,q) \text{ square free}}} \left|\mathcal{S}_2(x; a, q) - \delta_{2,a,q}x\right| \ll \frac{x}{\log^A x}$$

and for $y \leq x$

$$\sum_{q \leq y} \sum_{\substack{a \in \mathbb{Z}/q\mathbb{Z} \\ \gcd(a,q) \text{ square free}}} \left|\mathcal{S}_2(x; a, q) - \delta_{2,a,q}x\right|^2 \ll xy + x^{8/5}\log^5 x.$$

Warlimont in 1969 [41] proved some similar statements showing that if $y \leq x$

$$\sum_{q \leq y} \sum_{a \in (\mathbb{Z}/q\mathbb{Z})^*} \left|\mathcal{S}_2(x; a, q) - \delta_{2,q}x\right|^2 \ll \begin{cases} xy & \text{if } 1 \leq y \leq x^{\frac{1}{3}} \\ & \text{or } y \geq x^{1/3}\log^{10/3} x; \\ y^{\frac{4}{3}}x^{\frac{2}{3}+\epsilon} & \text{if } y = x^{\alpha},\ \frac{1}{2} \leq \alpha \leq 1. \end{cases}$$

To the author's knowledge, the above results have not been generalized to the case of $k$–free numbers.

We conclude this section mentioning the results of Suryanarayana [37] obtained in 1969 where he showed that the number of $k$-free integers $\leq x$ which are relatively prime to $m$ equals

$$\varphi(m)\delta_{k,m}x + O\left(\frac{\varphi(m)2^{\omega(m)}}{m}x^{1/k}\right),$$

saving a factor of $\frac{2^{\omega(m)}}{m}$ with respect to the direct estimate.

In the same paper Suryanarayana also shows that

$$\sum_{\substack{n\le x\\ \gcd(n,m)=1\\ n\ k\text{–free}}} n = \varphi(m)\delta_{k,m}\frac{x^2}{2} + O\left(\frac{\varphi(m)2^{\omega(m)}}{m}x^{1+1/k}\right)$$

and that

$$\sum_{n\le x}\sum_{\substack{d|n\\ (d,n/d)=1\\ d\ k\text{–free}}} d = \prod_{l}\left(1-\frac{1}{l^k}+\frac{1}{l^2+l}\right)\cdot\frac{x^2}{2} + O(x^{1+1/k}).$$

## 3. The general problem of $k$–freeness

If $f:\mathbb{N}\to\mathbb{Z}$ is any function, we define

$$\mathcal{S}_f^k(x) := \#\{n\le x \mid f(n) \text{ is } k\text{–free}\}.$$

We have the following clear identity:

$$\mathcal{S}_f^k(x) = \sum_{d=1}^{\infty}\mu(d)\#\{n\le x \mid d^k \mid f(n)\}$$

which is suggesting that if the probability that $f(n)$ is divisible by an integer $D$ is given by a number $\mathbb{P}_f(D)$, then one might expect to hold an asymptotic formula of the type

$$\mathcal{S}_f^k(x) \sim \prod_{l}\left(1-\mathbb{P}_f(l^k)\right)x.$$

This problems turns out to be quite challenging in general and the above formula is not often known to hold.

A natural variation to the problem is to restrict the argument of the function $f$ to prime values. Therefore we set:

$$\tilde{\mathcal{S}}_f^k(x) := \#\{p\le x \mid p \text{ is prime and } f(p) \text{ is } k\text{–free}\}.$$

We mention that a result of Mirsky of 1949 [25] deals with the case when $f(t)=t+a$ with $a\in\mathbb{Z}$. In this case for any constant $A$,

$$\tilde{\mathcal{S}}_f^k(x) := \#\{p\le x \mid p \text{ prime, } p+a \text{ is } k\text{–free}\} = \beta_a\pi(x) + O\left(\frac{x}{\log^A x}\right)$$

where $\pi(x)$ is the number of primes $p \leq x$ and

$$\beta_a = \prod_{l \text{ prime}} \left(1 - \frac{1}{l^k(l-1)}\right)$$

is the so called *Artin k–dimensional constant.*

We would like here to prove this statement with a much weaker error term. Indeed if $\pi(x; -a, q)$ counts the number of primes $p \leq x$, $p \equiv -a \bmod q$, then using some prime number Theorem for arithmetic progressions,

$$\begin{aligned}
\tilde{S}_f^k(x) &= \sum_{d \leq x^{1/k}} \mu(d)\pi(x; -a, d^k) \\
&= \sum_{d \leq \log x} \mu(d)\left(\frac{\pi(x)}{\varphi(d^k)} + O\left(\frac{x}{\log^3 x}\right)\right) + \left(\sum_{\substack{d \leq x^{1/k} \\ d > \log x}} \pi(x; -a, d^k)\right) \\
&= \beta_a \pi(x) + O\left(\sum_{d > \log x} \frac{\pi(x)}{\varphi(d^k)} + \frac{x}{\log^2 x} + \sum_{d > \log x} \frac{x}{d^k}\right) \\
&= \beta_a \pi(x) + O\left(\frac{x}{\log^2 x}\right)
\end{aligned}$$

where we used the clear estimate $\pi(x; -a, q) \ll x/q$. This concludes the proof.

The last result that we mention in this section is due in 1998 to Brüdern, Granville, Perelli, Vaughan and Wooley [8]. They consider $L^1$ and $L^2$ means of exponential sums supported on $k$–free numbers. As a corollary of various results, they show that

$$\#\{n \leq x \mid n = a^3 + b^3 + c^3, a, b, c \in \mathbb{N} \text{ and } n \text{ is square free}\} \gg x^{11/12}.$$

## 4. The case of polynomials in one variable

In this section we consider the case when $f(t) \in \mathbb{Z}[t]$ is a polynomial without multiple roots, primitive and such that the greatest common divisors of all the $f(n)$'s is $k$–free (to avoid pathologies like $f(t) = t(t+1)(t+2)(t+3)$ whose values are always divisible by 8).

This interesting case has also attracted several authors.

After Nagell in 1922 [28] proved that a polynomial of degree $\deg f$ less or equal then $k$ assumes infinitely many $k$–free values, the Italian mathematician Ricci in 1933 [36] proved that if

$$\delta_{f,k} = \prod_{l \text{ prime}} \left(1 - \frac{\varrho_f(l^k)}{l^k}\right)$$

where once again $\varrho_f(d)$ is the number of roots of $f$ in $\mathbb{Z}/d\mathbb{Z}$, then

$$\boxed{\mathcal{S}_f^k(x) \sim \delta_{f,k} x}$$

provided that $\deg f \leq k$. The asymptotic formula above is conjectured to hold also when $\deg f > k$ and this is the main open question in this part of the subject.

Here we would like to prove the statement of Ricci's Theorem and more precisely that for any $\epsilon > 0$

$$\boxed{\mathcal{S}_f^k(x) = x\delta_{f,k} + O\left(\frac{x}{\log^{1-\epsilon} x}\right) \qquad \text{if } \deg f \leq k.}$$

Indeed write $r$ for $\deg f$, let $z = \log x$ and $P(z)$ denote the product of all primes up to $z$, write

$$\begin{aligned}
\mathcal{S}_f^k(x) &= \sum_{n \leq x} \mu^{(k)}(f(n)) \\
&= \sum_{d \geq 1} \mu(d) \#\{n \leq x \mid d^k \mid f(n)\} \\
&= \sum_{d \mid P(z)} \mu(d) \#\{n \leq x \mid d^k \mid f(n)\} + O\left(\sum_{p > z} \#\{n \leq x \mid p^k \mid f(n)\}\right). \qquad (3)
\end{aligned}$$

Now note that

$$\#\{n \leq x \mid d^k \mid f(n)\} = \varrho_f(d^k)\left(\frac{x}{d^k} + O(1)\right).$$

Furthermore we have that

1) if $p^k \mid f(n)$, then $p \leq cx^{r/k}$ for a suitable constant $c = c(k, f)$ (Note that the hypothesis $r/k \leq 1$ will be crucial in the sequel);
2) $\varrho_f$ is a multiplicative function;
3) $\varrho_f(p^k)$ is uniformly bounded in term of $k$ and $f$ for all primes $p$. So there exists $\tilde{c} = \tilde{c}(k, f)$ such that $\varrho_f(p^k) \leq \tilde{c}$

Hence (3) equals:

$$\begin{aligned}
&\sum_{d \mid P(z)} \mu(d)\varrho_f(d^k)\left(\frac{x}{d^k} + O(1)\right) + O\left(\sum_{p > z} \frac{x}{p^k} + \sum_{p \leq cx^{r/k}} 1\right) \\
&= x \sum_{d \mid P(z)} \frac{\mu(d)\varrho_f(d^k)}{d^k} + O\left((1+\tilde{c})^{\pi(z)} + \frac{x}{z^{k-1}\log z} + \frac{x^{r/k}}{\log x}\right) \\
&= x \sum_{d=1}^{\infty} \frac{\mu(d)\varrho_f(d^k)}{d^k} + O\left(x \sum_{d > z} \frac{\tilde{c}^{\omega(d)}}{d^k} + \frac{x}{\log x \log\log x}\right) \\
&= x\delta_{f,k} + O\left(\frac{x}{\log^{1-\epsilon} x}\right),
\end{aligned}$$

where we used the estimate $\tilde{c}^{\omega(d)} = d^{\epsilon}$ and this completes the proof.

The next case, degree of $f$ equals $k+1$, had to wait until 1953, when Erdös [10] under this hypothesis showed that $\mathcal{S}_f^k(x) \to \infty$ as $x \to \infty$.

In 1976 Hooley [18], improving his results of 1967 [16,17], applies the large sieve to Ricci's proof showing that if $f$ is irreducible, then

$$\mathcal{S}_f^k(x) = x\delta_{f,k} + O\left(\frac{x}{\log^{k/(k+2)} x}\right) \qquad \text{if } \deg f = k+1.$$

The next conquest is due to Nair [29] who in 1976 proves that if $\lambda = \sqrt{2} - 1/2 = 0.9142\cdots$, and $f$ is irreducible, then

$$\mathcal{S}_f^k(x) = x\delta_{f,k} + O\left(\frac{x}{\log^{k-1} x}\right) \qquad \text{if } \deg f \geq \lambda k.$$

In two subsequent papers [21,30], the second in collaboration with Huxley, Nair improves his results proving that for some $\sigma = \sigma(k, f) > 1$ that can be explicitly computed, one has

$$\mathcal{S}_f^k(x) = x\delta_{f,k} + O\left(x^{1-\sigma}\right) \quad \text{if } k \geq \sqrt{2(\deg f)^2 + 1} - (\deg f + 1)/2.$$

Furthermore in the same paper Nair also proves that there exists a suitable positive $\alpha \leq 70/71$ such that

$$\mathcal{S}_f^k(x) = x\delta_{f,k} + O\left(x^{\alpha}\right) \qquad \text{if } \deg f = k+1 \geq 7.$$

Nair's approach was also employed by Hinz [14] in 1982 to prove generalizations of these results in the algebraic number fields settings.

More recently in 1998, Granville [12] proves that the *abc*–conjecture of Œsterle, Masser and Szpiro implies that if $f \in \mathbb{Z}[t]$ does not have repeated roots and if the greatest common divisor of all values $f(n)$ is square free, then the conjecture always holds for $f$ when $k = 2$:

$$\textbf{abc} \quad \Longrightarrow \quad \mathcal{S}_f^2(x) \sim \delta_{f,2}x.$$

Granville's statement is indeed more general and does not assume that the greatest common divisor of all values $f(n)$ is square free. Furthermore his paper [12] deals also with other problems involving square-free numbers.

The first ones to use the *abc*–conjecture to the problem of counting square free values of polynomials were Browkin, Filaseta, Greaves and Schinzel [7], who in 1995 proved that the *abc*–conjecture implies that the cyclotomic polynomial $\Phi_m(n)$ is square free for infinitely many values of $n$.

For a given integer $d$, we set $\tilde{\varrho}_f(d)$ to be the number of roots of $f(t)$ in $(\mathbb{Z}/d\mathbb{Z})^*$. Furthermore we define

$$\tilde{\delta}_{f,k} = \prod_{l \text{ prime}} \left(1 - \frac{\tilde{\varrho}_f(l^k)}{l^{k-1}(l-1)}\right).$$

It is reasonable to conjecture that

$$\tilde{\mathcal{S}}_f^k(x) \sim \tilde{\delta}_{f,k} \frac{x}{\log x}.$$

Indeed if the degree of $f$ is less or equal to $k$, this can be proven along the lines of Ricci's Theorem. However the above is considered to be in general a more difficult problem with respect to the previous one.

It is a result of Hooley [19] of 1977 that if $\deg f = k+1 \geq 55$ and $f$ is irreducible, then there exists a constant $\Delta_k > 0$ such that

$$\tilde{\mathcal{S}}_f^k(x) = \tilde{\delta}_{f,k} \frac{x}{logx} + O\left(\frac{x}{\log^{1+\Delta_k} x}\right) \quad \text{if } \deg f = k+1 \geq 55.$$

Hooley [20] later improved the above range for $k$ to $k \geq 41$, under some technical conditions. Nair [29,30] and later Huxley and Nair [21] also consider this problem.

## 5. The case of classical arithmetical functions

As we will see, the behavior of $\mathcal{S}_f^k$ in the case of classical arithmetical functions is rather different from the case of polynomials.

We recall that the Carmichael function $\lambda(n)$ is defined as the largest possible order of any element in the unit group of the residues modulo $n \geq 1$. More explicitly, for a prime power $p^\nu$, we have

$$\lambda(p^\nu) = \begin{cases} p^{\nu-1}(p-1), & \text{if } p \geq 3 \text{ or } \nu \leq 2; \\ 2^{\nu-2}, & \text{if } p = 2 \text{ and } \nu \geq 3 \end{cases}$$

and for arbitrary $n \geq 2$,

$$\lambda(n) = \text{lcm}(\lambda(p_1^{\nu_1}), \ldots, \lambda(p_s^{\nu_s})),$$

where $n = p_1^{\nu_1} \cdots p_s^{\nu_s}$ is the prime factorization of $n$. Clearly $\lambda(1) = 1$. Furthermore for every integer $n$ we have that $\lambda(n) \mid \varphi(n)$ and the equality holds when $n$ is a prime power.

Another cognate function is the order function: if $a \in \mathbb{Z} \setminus \{0, \pm 1\}$, then for each $n$ coprime to $a$, we define

$$\text{ord}_a(n) := \min\{e \in \mathbb{N} \mid a^e \equiv 1 \bmod n\}.$$

Clearly $\text{ord}_a(n) \mid \lambda(n)$. It is convenient to extend the definition of $\text{ord}_a(n)$ setting it to be equal to 0 when $\gcd(a, n) \neq 1$.

In [34], Saidak, Shparlinski and the author prove that

$$\mathcal{S}_\lambda^k(x) = (\kappa_k + o(1))\,\frac{x}{\log^{1-\alpha_k} x},$$

where

$$\kappa_k := \frac{2^{k+2}-1}{2^{k+2}-2}\cdot\frac{\eta_k}{e^{\gamma\alpha_k}\Gamma(\alpha_k)},\quad \alpha_k := \prod_{l \text{ prime}}\left(1-\frac{1}{l^{k-1}(l-1)}\right)$$

and $\eta_k$ is defined by

$$\eta_k := \lim_{T\to\infty}\frac{1}{\log^{\alpha_k} T}\prod_{\substack{l\le T\\ l-1 \text{ is } k\text{-free}}}\log\left(1+\frac{1}{l}+\ldots+\frac{1}{l^k}\right).$$

For example we have $k_2 = 0.80328\ldots$ and $\alpha_2 = 0.37395\ldots$.

As for the order function, in [35] the author proves that

$$\mathcal{S}_{\mathrm{ord}_a}(x) = \left(\iota_{a,k} + o(1)\right)\frac{x}{\log^{1-\beta_{a,k}} x}.$$

The constants $\iota_{a,k}$ and $\beta_{a,k}$ have a complicated definition. However in the case when $a$ is square free and $k \ge 3$, the definition for $\beta_{a,k}$ simplifies in

$$\beta_{a,k} := \left[\prod_l\left(1-\frac{1}{l^{k-2}(l^2-1)}\right)\right]\cdot\left[1-\frac{1}{2}\prod_{l\mid[2,a]}\frac{1}{1-l^{k-2}(l^2-1)}\right].$$

The proof of both results follows the same framework using a classical statement, due to Wirsing [42]:

**Theorem.** *Assume that $g(n)$ is multiplicative, $0 < g(p^\nu) \le c_5 c_6^\nu$, $c_6 < 2$ and*

$$\sum_{p\le x} g(p) = (1+o(1))\tau\pi(x)$$

*for some $\tau \ne 0$. Let $\gamma$ denote the Euler constant, and $\Gamma$ the gamma-function. Then*

$$\sum_{n\le x} g(n) \sim \frac{1}{e^{\gamma\tau}\Gamma(\tau)}\frac{x}{\log x}\prod_{l\le x}\sum_{\nu=0}^{\infty}\frac{g(l^\nu)}{l^\nu}.$$

It is immediate to check that both

$$\mu^{(k)}(\lambda(n)) \quad\text{and}\quad \mu^{(k)}(\mathrm{ord}_a(n))$$

are multiplicative functions. In order to apply Wirsing's Theorem, one needs to find an asymptotic formula for

$$\tilde{\mathcal{S}}_\lambda^k(x) = \#\{p \le x \mid p-1 \text{ is } k\text{-free}\}$$

in the case of λ, and for

$$\tilde{\mathcal{S}}^k_{\mathrm{ord}_a}(x) = \#\{p \le x \mid p \nmid a,\ \mathrm{ord}_p(a) \text{ is } k\text{–free}\}$$

in the case of $\mathrm{ord}_a$.

The first of these asymptotic formulas is due to Mirsky [25] and we already referred to it in the last part of Section 3.. For any constant $A > 0$ we have

$$\tilde{\mathcal{S}}^k_\lambda(x) = \alpha_k \pi(x) + O\left(\frac{x}{\log^A x}\right).$$

The second asymptotic formula is proven in [35] by using the Chebotarev density Theorem in an unconditional version:

$$\tilde{\mathcal{S}}^k_{\mathrm{ord}_a}(x) = \beta_{a,k}\pi(x) + O\left(\frac{x}{\log^\sigma x}\right),$$

where $\sigma = \sigma(a, k)$ can be explicitly computed.

In general if $f$ in any function such that $\mu^{(k)}(f(n))$ is multiplicative, then Wirsing's Theorem gives that

$$\tilde{\mathcal{S}}^k_f(x) \sim \tau_{f,k}\pi(x) \quad \Longrightarrow \quad \mathcal{S}^k_f(x) \sim \frac{\pi(x)}{e^{\gamma\tau_{f,k}}\Gamma(\tau_{f,k})} \prod_{p \le x} \sum_{\nu=0}^{\infty} \frac{\mu^{(k)}(p^\nu)}{p^\nu}.$$

Unfortunately Wirsing's Theorem does not provide any error term.

The last classical function that we will discuss is the Euler function. It has a rather different behavior with respect to the others. Indeed Banks and the author in [5] showed that for a fixed integer $k \ge 3$, the asymptotic relation

$$\mathcal{S}^k_\varphi(x) = \frac{3\alpha_k}{2} \frac{x\,(\log\log x)^{k-2}}{(k-2)!\,\log x}\left(1 + O_k\left(\frac{(\log\log\log x)^{2(k+1)^{2k-4}-1}}{(\log\log x)^{1-1/k}}\right)\right)$$

holds as $x \to \infty$ with the constant $\alpha_k$ defined by

$$\alpha_k := \frac{1}{2^{k-1}} \prod_{l>2}\left(1 - \frac{1}{l^{k-1}} \sum_{i=0}^{k-2}\sum_{j=0}^{k-2-i} \binom{k-1}{i}\binom{k-1+j}{j}\frac{(l-2)^j}{(l-1)^{i+j+1}}\right).$$

The above statement reminds Landau's Theorem for the number of integers up to $x$ with at most $k-1$ prime factors. Indeed if $\varphi(n)$ is $k$–free, then $\omega(n) \le k-1$.

## 6. $k$–free numbers with restricted digits

Let $g \in \mathbb{N}$, $\mathcal{D} \subset \{0, 1, \dots, g-1\}$ be a subset with $d$ elements and let $e$ be the number of elements in $\mathcal{D}$ which are coprime with $g$. Furthermore denote by $\mathcal{A}_g(\mathcal{D})$ the set of those integers which are coprime with $g$ and such that all their digits in base $g$ are in $\mathcal{D}$.

It is natural to consider the counting function

$$\mathcal{S}^k_{\mathcal{D},g}(x) = \#\{n \le x \mid n \text{ is } k\text{–free and } n \in A_g(\mathcal{D})\}.$$

Banks and Shparlinski [4] prove that there exists a constant $c > 0$ depending on $\mathcal{D}$ and $k$ such that if $x = g^m$ and $m \to \infty$, then

$$\mathcal{S}^k_{\mathcal{D},g}(x) = \delta_{\mathcal{D},g} x^{\log_g d} + O\left(\frac{x^{\log_g d}}{\exp(c\sqrt{\log x})}\right),$$

where

$$\delta_{\mathcal{D},g} = \frac{e}{\zeta(k)d}\prod_{l|g}\left(1-\frac{1}{l^k}\right)^{-1}.$$

Furthermore if $r$ is an integer, $\mathcal{B}_r$ denotes the set of those odd integers having exactly $r$ ones in their binary expansion and

$$\mathcal{S}^k_{\mathcal{B}_r}(x) = \#\{n \le x \mid n \text{ is } k\text{–free and } n \in \mathcal{B}_r\}.$$

Then there exists a constant $d > 0$ such that, if $x = 2^m$ and $m \to \infty$, then

$$\mathcal{S}^k_{\mathcal{B}_r}(x) = \frac{\binom{m-1}{k-1}}{\zeta(k)(1-2^{-k})}\left(1 + O\left(\frac{1}{\exp(d\delta\sqrt{\log x})}\right)\right),$$

provided that $r/m \in [(\log 2 + \delta)/k, 1/2]$.

## 7. Conclusion

Many variations of the results in the literature could be considered and would provide a good project for young mathematicians. We list a few here. Some might be easier, some might be very difficult, others might already be known.

(1) Find an asymptotic formula for $\mathcal{S}^k_f(x)$ when $f$ is some other classical arithmetic functions: e.g. $f = \sigma, \tau, \omega, \Omega$;
(2) Find an asymptotic formula for $\mathcal{S}^k_f(x)$ when $f(n)$ is the $n$–th coefficient of a modular form;
(3) Find an asymptotic formula for $\mathcal{S}^k_{\mathrm{ind}_a}(x)$, where the index function is defined as $\mathrm{ind}_a(n) = \varphi(n)/\operatorname{ord}_a(n)$. Some partial results already appeared in [33];
(4) Find an asymptotic formula for the values of an arithmetic function on numbers in arithmetic progressions;

(5) Given a finitely generated subgroup $\Gamma \subset \mathbb{Q}^*$, we know that for all but finitely many primes $p$ we can consider the reduction $\langle \Gamma \bmod p \rangle \subset (\mathbb{Z}/p\mathbb{Z})^*$. Furthermore there exists an integer $M_\Gamma$ such that if $\gcd(n, M_\Gamma) = 1$, then the reduction $\langle \Gamma \bmod n \rangle$ is a well defined subgroup of $(\mathbb{Z}/n\mathbb{Z})^*$. Find an asymptotic formula for the number of integers $n$ up to $x$ such that $(n, M_\Gamma) = 1$ and $\#\langle \Gamma \bmod n \rangle$ is $k$–free.

We realize that this survey is far from complete. For example we completely left out important topics like $k$–free values of multivariate polynomials or binomial coefficients, gaps between $k$–free numbers and surely many others.

## Acknowledgments

This paper was written while the author was visiting the Harish Chandra Research Institute in Allahabad. He would like to thank his hosts for financial support and hospitality. The author was also supported by GNSAGA from INDAM.

## References

[1] Axer, A. *Über einige Grenzwertsätze.* S.-B. Math.-Natur. K1. Akad. Wiss. Wien (2a) **120** (1911) 1253–1298

[2] Baker, R. C. and Pintz, J. The distribution of squarefree numbers. *Acta Arith.* **46** (1985) no. 1 73–79

[3] Balasubramanian, R. and Ramachandra, K. Some problems of analytic number theory II. *Studia Sci. Math. Hungar.* **14** (1979) no. 1-3 193–202 (1982)

[4] Balasubramanian, R. and Ramachandra, K. *On square-free numbers*. Proceedings of the Ramanujan Centennial International Conference (Annamalainagar, 1987), 27–30, RMS Publ., 1, Ramanujan Math. Soc., Annamalainagar (1988)

[5] Banks, W. D. and Pappalardi, F. *Powerfree Values of the Euler Function*, Preprint (2003)

[6] Banks, W. D. and Shparlinski, I. Arithmetic Properties of numbers with restricted digits. *Acta Arith.* (to appear)

[7] Browkin, J., Filaseta, M., Greaves, G., and Schinzel, A. Squarefree values of polynomials and the *abc*-conjecture. Sieve methods, exponential sums, and their applications in number theory (Cardiff, 1995), 65–85, *London Math. Soc. Lecture Note Ser.*, 237, Cambridge Univ. Press, Cambridge, (1997)

[8] Brüdern, J., Granville, A., Perelli, A., Vaughan, R. C. and Wooley, T. D. On the exponential sum over $k$-free numbers. *R. Soc. Lond. Philos. Trans. Ser. A Math. Phys. Eng. Sci.* **356** (1998), no. 1738, 739–761

[9] Cohen, E. and Robinson, R. L. On the distribution of the $k$-free integers in residue classes. *Acta Arith.* **8**, (1962/1963) 283–293. errata, ibid. **10** (1964/65) 443

[10] Erdös, P. Arithmetical properties of polynomials. *J. London Math. Soc.* **28** (1953) 416–425

[11] Graham, S. W. The distribution of squarefree numbers. *J. London Math. Soc.* (2) **24** (1981), no. 1, 54–64

[12] Granville, A. $ABC$ allows us to count squarefrees. *Internat. Math. Res. Notices* **19** (1998), 991–1009

[13] Graham, S. W. and Pintz, J. The distribution of $r$-free numbers. *Acta Math. Hungar.* **53** (1989), no. 1–2, 213–236

[14] Hinz, J. G. Potenzfreie Werte von Polynomen in algebraischen Zahlkörpern. *J. Reine Angew. Math.* **332** (1982) 134–150

[15] Hooley, C. A note on square-free numbers in arithmetic progressions. *Bull. London Math. Soc.* **7** (1975) 133–138

[16] Hooley, C. *On the power free values of polynomials.* Mathematika **14** (1967) 21–26

[17] Hooley, C. On the square-free values of cubic polynomials. *J. Reine Angew. Math.* **229** (1968) 147–154

[18] Hooley, C. Applications of sieve methods to the theory of numbers. *Cambridge Tracts in Mathematics*, No. 70. Cambridge University Press, Cambridge-New York-Melbourne, (1976)

[19] Hooley, C. On power-free numbers and polynomials I. *J. Reine Angew. Math.* **293/294** (1977) 67–85

[20] Hooley, C. On power-free numbers and polynomials II. *J. Reine Angew. Math.* **295** (1977), 1–21. errata ibid. **299/300** (1978) 433

[21] Huxley, M. N. and Nair, M. Power free values of polynomials III. *Proc. London Math. Soc.* (3) **41** (1980), no. 1, 66–82

[22] Jia, Chao Hua, The distribution of squarefree numbers. *Beijing Daxue Xuebao* (1987), no. 3, 21–27

[23] Jia, Chao Hua. The distribution of square-free numbers. *Sci. China Ser. A* **36** (1993), no. 2, 154–169

[24] Li, Hong Ze The distribution of $k$th-power-free numbers. *Chinese Quart. J. Math.* **3** (1988), no. 4, 10–16

[25] Mirsky, L. The number of representations of an integer as the sum of a prime and a $k$-free integer. *Amer. Math. Monthly* **56** (1949) 17–19

[26] McCurley, K. *The distribution of r-free integers in arithmetic progressions.* Topics in analytic number theory (Austin, Tex., 1982), 304–312, Univ. Texas Press, Austin, TX, (1985)

[27] Montgomery, H. L. and Vaughan, R. C. *The distribution of squarefree numbers.* Recent progress in analytic number theory, Vol. 1 (Durham, 1979) 247–256 Academic Press, London-New York, (1981)

[28] Nagell, T. Zur Arithmetik der polynome. *Abhandl. Math. Sem. Hamburg* **1** (1922) 179–194

[29] Nair, M. *Power free values of polynomials.* Mathematika **23** (1976), no. 2, 159–183

[30] Nair, M. Power free values of polynomials, II. *Proc. London Math. Soc.* (3) **38** (1979), no. 2, 353–368

[31] Orr, R. Remainder estimates for squarefree integers in arithmetic progression. *J. Number Theory* **3** (1971) 474–497

[32] Pappalardi, F. Squarefree Values of the Order Function. *New York Journal of Math.* (to appear)

[33] Pappalardi, F. *On Hooley's Theorem with weights.* Rend. Sem. Mat. Polit. Univ. Torino **53** (1995), no. 4 375–388

[34] Pappalardi, F., Saidak, F. and Shparlinski, I. E., Squarefree Values of the Carmichael Function. *J. Num. Theory* **103** (2003) 122–131

[35] Prachar, K. Über die kleinste quadratfreie Zahl einer arithmetischen Reihe. *Monatsh. Math.* **62** (1958) 173–176

[36] Ricci, G. *Ricerche aritmetiche sui polinomi.* Rend. Circ. Mat. Palermo **57** (1933) 433–475

[37] Suryanarayana, D. The number and sum of $k$-free intergers $\leq x$ which are prime to $n$. *Indian J. Math.* **11** (1969) 131–139

[38] Vaidya, A. M. On the changes of sign of a certain error function connected with $k$-free integers. *J. Indian Math. Soc.* (N.S.) **32** (1968) 105–111

[39] Yao, Qi A note on the distribution of $k$th power free numbers. *Shanghai Keji Daxue Xuebao* (1984), no. 4, 9–15

[40] Walfisz, A. *Weylsche Exponentialsummen in der neueren Zahlentheorie.* Mathematische Forschungsberichte, XV. VEB Deutscher Verlag der Wissenschaften, Berlin (1963)

[41] Warlimont, R. On squarefree numbers in arthmetic progressions. *Monatsh. Math.* **73** (1969) 433–448

[42] Wirsing, E. Das asymptotische Verhalten von Summen über multiplikative Funktionen. *Math. Ann.* **143** (1961) 75–102

Proc. Int. Conf. – Number Theory No. 1, 2004, pp. 89–95.

# Automorphic Forms and Dirichlet Series

S. Raghavan

(Retired Professor of Mathematics, TIFR, Mumbai), Flat S-1, Suryakantha, (New) 26, Second Main Road, C.I.T. Colony, Mylapore, Chennai 600 004
e-mail : ragrad@sancharnet.in

A part of Hecke's correspondence (1936) between *modular forms* $f$ of *weight* $k$ and 'multiplier' $c$ for the group $G(\lambda)$ generated in SL(2;**R**) by $\begin{pmatrix} 1 & \lambda \\ 0 & 1 \end{pmatrix}$ and $\begin{pmatrix} 0 & 1 \\ -1 & 0 \end{pmatrix}$ on the one hand and Dirichlet series associated on the other hand to $f$ in a specific manner is the *equivalence* of the following, namely

(i) $f(-1/z) = c(-iz)^k f(z)$ for a modular form $f$ of type $(\lambda, k, c)$ on $G(\lambda)$ with the Fourier expansion of $f$ given by $f(z) = \sum_{n=1}^{\infty} a_n e^{2\pi i n z/\lambda}$ holomorphic for $z \in H$, the complex upper half-plane and

(ii) $\Phi(s) := (2\pi/\lambda)^{-s}\Gamma(s)\sum_{n=1}^{\infty} a_n n^{-s}$ is an entire function of $s$ in **C**, bounded in vertical strips in **C** and has the functional equation $\Phi(k-s) = c\Phi(s)$. For the equivalence of (i) and (ii), a crucial role is played by the usual Mellin transform. In 1967,such a correspondence between (holomorphic) modular forms for congruence subgroups of G(1) = SL(2,**Z**) and Dirichlet series associated to them as above was established in A. Weil's important paper [12]; a single functional equation as in (ii) now gets replaced by a host of such equations under '$s$ to $k-s$' involving also Dirichlet series with coefficients $a_n$ of $f$ 'twisted' by values of Dirichlet characters. Meanwhile, in the forties (of the twentieth century), the theory of the Hecke correspondence as above got 'liberated' from what Weil referred to as a 'ghetto' of a requirement on $f$ to be holomorphic on $H$, thanks to Maass' innovative correspondence between real analytic (non-holomorphic) automorphic forms on $H$ occurring as eigenfunctions of the Laplacian on $H$ and associated (pairs of) Dirichlet series. Maass' paper [8] succeeded in extending Hecke's correspondence above to cover (real analytic) automorphic forms on $k$-dimensional hyperbolic space $H^k$ and associated Dirichlet series with coefficients involving 'twists' by spherical harmonic functions in $k$ variables. The possible indispensability of the spectral theory in $L^2(H/SL(2,\mathbf{Z}))$ for the Laplacian and similar objects became quite evident thereby,in the study of Hecke's correspondence between modular forms and Dirichlet series. The ensuing well-known work of Jacquet and Langlands established the deep and unmistakable link between the Hecke-Maass-Weil correspondence and representation theory; irreducible representations of GL(2) took over from automorphic forms, holomorphic or otherwise. This essentially expository article will be concerned mainly with an analogue of Hecke's correspondence for modular forms in several variables such as Siegel modular forms and Hermitian modular forms (especially of degree 2).

Let $\mathcal{S}_n := \{Z = {}^tZ \in M_n(C)$ with Im $(Z) := (1/2i)(Z - \overline{Z}) > 0\}$, the Siegel upper half-plane of genus $n$ (so that $\mathcal{S}_1$ is just the complex upper half-plane $H$) and let $\mathcal{H}_n := \{Z \in M_n(C)$with $\mathrm{Im}(Z) := (1/2i)(Z - \overline{{}^tZ}) > 0\}$, the Hermitian upper half-plane of degree $n$. These are homogeneous spaces acted upon discontinuously by the Siegel modular group

$$\Gamma_n := Sp(n, Z)$$

$$= \left\{\begin{pmatrix} A & B \\ C & D \end{pmatrix} \in M_{2n}(Z) | A^tD - B^tC = E_n, \quad A^tB - B^tA = C^tD - D^tC = 0\right\}$$

and respectively the Hermitian modular group

$$\mathcal{H}_n := \left\{\begin{pmatrix} A & B \\ C & D \end{pmatrix} \in M_{2n}(K) | A^t\overline{D} - B^t\overline{C} = E_n \quad \text{and} \quad A^t\overline{B} - B^t\overline{A} = 0 \right.$$

$$\left. = C^t\overline{D} - D^t\overline{C}\right\}$$

for a given imaginary quadratic field $K$ over **Q**. These two cases have been christened respectively as the 'case SP' and the 'case SU' by Shimura [11]. For a matrix $P$, its transpose is ${}^tP$ and its (complex) conjugate is $\overline{P}$; $E_n$ and 0 denote the $(n, n)$ identity and zero matrix respectively above.

A complex-valued function $f$ holomorphic in $\mathcal{S}_n$ is called a *Siegel modular form* of *degree n* and *weight k* if, for every $M = \begin{pmatrix} A & B \\ C & D \end{pmatrix} \in \Gamma_n$, we have

$$(f|M)(Z) := f((AZ + B)(CZ + D)^{-1}) \det(CZ + D)^{-k} = f(Z) \qquad (*)$$

while, for $n = 1$, $f$ is bounded in any (arbitrary) domain $\{z = x+iy \in H | y \geq y_0 > 0\}$.

For $n > 1$, it was shown by Koecher that the 'boundedness' property of $f$ (which has to be imposed in the case $n = 1$ for the definition of a modular form $f$) is an automatic consequence of (*) above holding for all $M \in \Gamma_n$. For a (Siegel) modular form $f$ of degree $n$ and weight $k$ ,we have the Fourier expansion $f(Z) = \sum_{T \geq 0} a(T)e^{2\pi i tr(TZ)}$ where *tr* stands for trace and the summation is over all non-negative definite $(n, n)$ matrices $T = (t_{ij})$ which are 'half-integral' (i.e. with $t_{ii}$ and $2t_{ij}$ in **Z**). The coefficients $a(T)$ have 'polynomial growth' in the entries of $T$; it is not hard to show that $a(T) \ll \det T^k$ (i.e. $\leq$ const. $\det T^k$). We also note that for $U$ in GL($n$, **Z**) and *nk* even, $a({}^tUTU) = a(T)$ which thus depends only on $\{T\}$, the 'GL(n,**Z**)-equivalence class' of $T$ (under the natural action taking $T$ to $T[U] := {}^tUTU$ for $U$ in GL($n$,**Z**)). If $a(T) \neq 0 \Rightarrow T > 0$, then $f$ is called a *cusp form* of degree $n$ (as we shall indeed require $f$ to be, in the sequel). The easily proved boundedness of (det (Im $(Z)))^{k/2} f(Z)$ on $\Gamma_n \backslash \mathcal{S}_n$ implies that for a cusp form $f, a(T) \ll (\det T)^{k/2}$. If one therefore associates, to the given $f$, the Dirichlet series $\sum_{\{T\}} a(T)(\det T)^{-s}$, it converges for $\mathrm{Re}(s) > 1 + (k + n^2 + n)/2$ and makes perfect sense therein, as a (holomorphic) function of the complex variable $s$ (the summation over $\{T\}$ being over the $\mathrm{GL}(n, \mathbf{Z})$-equivalence classes of $(n, n)$ half-integral $T > 0$). Multiplying such a series by $\gamma(s)$ made up mainly from appropriate gamma factors, Koecher's paper [6] sought, for cusp forms $f$ of degree $n$ and weight $k$, the functional equation

$$R_n(f; s) = (-1)^{nk/2} R_n(f; k - s) \quad \text{for} \quad R_n(f; s) := \gamma(s) \sum_{\{T\}} \frac{a(T)}{e(T)} (\det T)^{-s}$$

where the summation over $\{T\}$ is as already mentioned and $e(T) := \#\{U \in GL(n, \mathcal{Z}) | {}^t UTU = T\}$. The work of Arakawa [1] employing Klingen's generalized Eisenstein series, captured the conjectured functional equation for $R_n(f; s)$; indeed, $R_n(f; s)/\gamma(s)$ is known as the Koecher-Maass series. In the case involving the Epstein-Koecher zetafunction associated (for $n = 2$) to an $(m, m)$ integral $S > 0$ with $\det S = 1$ (and $m > 2$), a beautiful proof for the corresponding functional equation was given by Professor Carl Ludwig Siegel in a communication (in September 1961) to Professor Raghavan Narasimhan who kindly showed it to the present author.

The 'converse theorem' recovering (for $f$) the property as a modular form as (purely) a consequence of the functional equation(s) involved was still missing; more variables and (correspondingly) more functional equations seemed imperative, as was revealed through the work of K. Imai [4]. In the sequel, we shall try to describe her work, incorporating modifications needed to cover the case of Hermitian cusp forms of degree 2. It should be mentioned that the question of generalizing Hecke's correspondence to the case of Siegel cusp forms of degree $n$ (for congruence subgroups) has been tackled in Weissauer's Preprint [13] on "Siegel modular forms and Dirichlet series"; featuring therein are functional equations under '$s$ to $k-s$' for $L$-series associated to Siegel cusp forms but now with 'twists' by Maass' Groessencharakteren on $\mathrm{GL}(n,\mathbf{R})$. With regard to the Koecher-Maass series, we may cite the following remark from Boecherer's Preprint [2]: "Die analytische Eigenschaften dieser Reihe sind weigchend bekannt (nach Maass-Maryland Lectures), allerdings sind die Fragen nach einer Residuenformel sowie das Umkehrproblem bis heute aktuell geblieben".

The work of Imai highlights the inevitable role, in regard to 'converse theorems' as above, played by the spectral theory for the Laplacian $\Delta$ in $\mathrm{L}^2(\Gamma_1 \backslash \mathcal{S}_1)$. Let $\mathrm{S}\wp_2$ denote the space of (2,2) real symmetric positive definite matrices of determinant 1. For $Z \in \mathcal{S}_2$ and $Y = \mathrm{Im}\ (Z)$, we can write $Y = t^{1/2} W$, with $W \in \mathrm{S}\wp_2$ and $t = \det Y$. Again, $W$ in $\mathrm{S}\wp_2$ is of the form $\begin{pmatrix} y^{-1} & -xy^{-1} \\ -xy^{-1} & (x^2 + y^2)/y \end{pmatrix}$ leading to the correspondence that takes $W$ to $z = x + iy \in \mathcal{S}_1$. The transformation taking $z$ to $M\langle z \rangle := (az + b)(cz + d)^{-1}$ for $M = \begin{pmatrix} a & b \\ c & d \end{pmatrix} \in \mathrm{SL}(2,\mathbf{R})$ corresponds to the bijection taking $W$ to $W[M^{-1}] := {}^t M^{-1} W M^{-1}$ of $S\wp_2$. The map '$z$ to $-z^{-1}$' corresponding to $M = \begin{pmatrix} 0 & -1 \\ 1 & 0 \end{pmatrix} \in \Gamma_1$ is the counterpart of '$W$ to $W^{-1}$', in view of the special relation

$W\begin{bmatrix}0 & -1\\ 1 & 0\end{bmatrix} = W^{-1}$ (See the last formula on page 929 of Imai[4]). Let us note that

$$S\wp h_2 := \{W = {}^t\overline{W} > 0 |\det W = 1\}$$

$$= \left\{ W = \begin{pmatrix} y^{-1} & -w/y \\ -\overline{w}/y & (y^2+|w|^2)/y \end{pmatrix} \middle| y > 0,\, w \in C \right\}$$

is in bijection with the 3-dimensional hyperbolic space $H^3 = \{\zeta = w + yj | w \in \mathbf{C}, y > 0\}$ under '$W \leftrightarrow \zeta$', though, however, $W\begin{bmatrix}0 & -1\\ 1 & 0\end{bmatrix} = {}^tW^{-1}$ (and not $W^{-1}$ as in the case of $S\wp_2$). For $\zeta = w + yj \in H^3$,the conjugate is $\tilde{\zeta} := \overline{w} - yj$ and the norm $N(\zeta) := \zeta\tilde{\zeta}$; for '$W \leftrightarrow \zeta$', we have $W^{-1} \leftrightarrow -\zeta^{-1}(= -\tilde{\zeta}/N(\zeta))$ and '$-\tilde{\zeta} \leftrightarrow {}^tW$'.

Let $f$ be a given Siegel cusp form of degree 2 and weight $k$. For $Z \in \mathcal{S}_2$ and $Y = \mathrm{Im}(Z)$, writing $Y = t^{1/2}W$ with $W \in S\wp_2$ and $t = \det Y$ as above and smoothening out by integration of $f(iY)$ with respect to the parameter $t$ (and cutting down dimensions by 1 thereby), one associates to $f$, via the theory of the Mellin transform, the function $\mathcal{M}f$ by

$$\mathcal{M}f(W; s) = (\mathcal{M}f(z; s)) := \int_0^\infty f(it^{1/2}W)t^s\frac{dt}{t} \quad \text{(for large enough} \quad \sigma := \mathrm{Re}(s)).$$

The next step is to estimate $f(iY)$ uniformly for $Y \geq Y_0 > 0$ with $Z$ in $\mathcal{S}_2$; assuming $Y$ to be reduced in the sense of Minkowski (as indeed justifiable), there exists a constant $c$ such that $Y - cY_0 > 0$ where $Y_0$ is diagonal with the same diagonal elements as $Y$; *thus* $Y \geq c\gamma(Y_0)E_2, tr(TY) \geq 2c\gamma(Y_0)(\det T)^{1/2}$ which together with $a(T) \ll (t_{11}t_{22})^\alpha$ for suitable $\alpha$ independent of $T$ as known, yield the estimate

$$|f(iY)| < \sum_{t=1}^{\infty} t^\alpha t^2(4t+1)e^{-4\pi c\gamma(Y_0)\sqrt{t}}.$$

On the other hand, given numbers $b(T)$ in $\mathbf{C}$ for (2, 2) half-integral $T = (t_{ij}) > 0$ with $b(T) \ll (t_{11}t_{22})^\alpha$ and $b(T[U]) = (\det U)^k b(T)$ for every $U$ in GL($n$,$\mathbf{Z}$), the series $g(Z) := \sum_{T>0} b(T)e^{2itr(TZ)\pi}$ converges absolutely, uniformly in any domain $Y \geq Y_1 > 0$ and is bounded; further,

$$g(iY) < ((\det Y)^{-(\alpha+4)} + (\det Y)^{-(\alpha+3)})e^{-4\pi\gamma(Y_1)\sqrt{\det Y}}.$$

**Proposition 1.** *For* $g(Z) := \sum_{T>0} b(T)e^{2itr(TZ)\pi}$ *as above and* $\mathcal{M}g$ *associated to* $g$ *likewise,* $W \in S\wp_2$ *and* $\sigma > \alpha + 4$, *we have*

$$\mathcal{M}g(W; \sigma + i\tau) \ll \gamma(Y_0)^{-2(\sigma-\alpha-3)}((\gamma(Y_0)^2\Gamma(2(\sigma-\alpha-4)) + \Gamma(2(\sigma-\alpha-3)))$$

*and so* $\mathcal{M}g$ *is bounded on* $S\wp_2$.

*Using the boundedness of* $\mathcal{M}g$ *in vertical* $s$*-strips and appropriate functional equations as well as the spectral expansion for* $\mathcal{M}g$ *in* $L^2(H/SL(2,\mathbf{Z}))$, *one tries to establish via Mellin inversion, automorphic properties, if any, for* $g(Z)$. *Actually, we have*

$$g(i\sqrt{t}W) = \frac{1}{2i\pi}\int_{\mathrm{Re}\ s=c'} \mathcal{M}g(W;s)t^{-s}\mathrm{d}s.$$

Reverting to the Siegel cusp form $f$, we have, in $L^2(H/\mathrm{SL}(2,\mathbf{Z}))$, the Roelcke-Selberg spectral decomposition

$$\mathcal{M}f(W;s) = (\mathcal{M}f(z;s)) = \sum_{j=0}^{\infty}\Phi(s,j)\nu_j(z) + \frac{1}{4i\pi}\int_{\mathrm{Re}\ u=1/2}\Phi(s,\rho)E_\rho(z)\mathrm{d}u.$$

Here $E_\rho(z) := \frac{1}{2}\sum_{\sigma\in\Gamma/\Gamma_\infty}\mathrm{Im}(\sigma\langle z\rangle)^\rho$ is the Eisenstein series defined for Re $\rho > 1$, with $\Gamma := \mathrm{SL}(2,Z)$, $\Gamma_\infty := \left\{\begin{pmatrix} a & b\\ 0 & d\end{pmatrix}\in\Gamma\right\}$, $\Delta\mathrm{E}_\rho = \rho(\rho-1)\mathrm{E}_\rho$, $\nu_0 = \sqrt{3/\pi}$, $\{\nu_j\}$ for $j = 1,\cdots,\infty$ is an orthonormal basis of eigenfunctions of $\Delta := y^2\left(\frac{\partial^2}{\partial x^2}+\frac{\partial^2}{\partial y^2}\right)$ (with 0 as the constant term in the Fourier expansion with respect to $x = \mathrm{Re}(z)$ and $\lambda_j$ as corresponding eigenvalue), $\Phi(s,\rho) := \langle Mf(\bullet,s),\mathrm{E}_\rho\rangle$, $\Phi(s,j) := \langle Mf(\bullet,s),\nu_j\rangle$ and $\langle\phi,\psi\rangle := \int\phi\overline{\psi}\mathrm{d}v$, the integration being over $\mathcal{F} := H/\mathrm{SL}(2,\mathbf{Z})$ with respect to $dv := y^{-2}\mathrm{d}x\mathrm{d}y$ for $\mathrm{L}_2$ functions $\phi,\psi$. The Eisenstein series $E_\rho$ is analytically continuable to $\mathrm{Re}(\rho)\geq 1/2$, with a simple pole at $\rho = 1$. Its Fourier expansion (with respect to Re $z$) is valid even for $\mathrm{Re}(\rho) = 1/2$; the Fourier coefficients involve the Bessel function $K_{\rho-1/2}$ and $E_\rho$ admits a functional equation under '$\rho$ to $1-\rho$'. By the way, no explicit examples of $\nu_j$ (or $\lambda_j$) are perhaps known as yet. (In the case of $\mathcal{H}_2$, $W$ in $S\wp h_2$ corresponds to $\zeta = w + yj$ in $H^3$, $-\tilde{\zeta} \leftrightarrow {}^tW$ and we have an Eisenstein series $\mathrm{E}_\rho(\varsigma)$ defined for $\mathrm{Re}(\rho) > 2$ and admitting analytic continuation to $\mathrm{Re}(\rho)\geq 1$ with a simple pole at $\rho = 2$ and a functional equation under '$\rho$ to $2-\rho$'. Recent work of Hoffstein and coauthors involving Siegel zeros raises hopes to improve (our) earlier estimates for Fourier coefficients of Maass cusp forms on $H^3$).

**Proposition 2.** *For* $(s,\rho)\in\mathcal{C}^2$ *with* $\{Re\ s > (Re\ \rho + k + 10)/2, Re\ r > 1\}$ *or* $\{Re\ s > (Re\ \rho + k + 8)/2, Re\ r > 2\}$, *we have, with* $e(T)$ *defined above and* $\Delta\overline{\nu}_j = \lambda_j\overline{\nu}_j$,

$$\Phi(s,\rho) = 2\cdot(2\pi)^{-2s}\sqrt{\pi}\Gamma(s-\overline{\rho}/2)\Gamma(s+(\overline{\rho}-1)/2)$$
$$\times\sum_{\{T\}\bmod P:=\left\{\begin{pmatrix} a & b\\ 0 & c\end{pmatrix}\right\}\in GL(2,\mathbf{Z})\}} a(T)t_{11}^{-\overline{\rho}}(\det T)^{-s+\overline{\rho/2}}$$

*and*

$$\Phi(s,j) = 2^{1-2s}\pi^{(1/2)-2s}\Gamma\left(s-\frac{1+\sqrt{1+4\lambda_j}}{4}\right)\Gamma\left(s-\frac{1-\sqrt{1+4\lambda_j}}{4}\right)$$
$$\times\sum_{0<T\bmod\mathrm{GL}(2,\mathbf{Z})}\frac{a(T)}{e(T)}\frac{\overline{\nu}_j(\det T^{-1/2}T)}{(\det T)^s}.$$

Further, we have the functional equations $\Phi(s,\rho) = \frac{\sqrt{\pi}\Gamma(\rho-1/2)\zeta(2\rho-1)}{\Gamma(\rho)\zeta(2\rho)}\Phi(s,1-\rho)$ and $\Phi(k-s,j) = (-1)^k\Phi(s,j)$ for $j\in\{0,1,2,\dots\}$ with $\zeta$, $\Gamma$ denoting respectively Riemann's zeta and Euler's gamma function.

(The proof for the first-mentioned formula for $\Phi(s, \rho)$, for example, involves the consideration of a majorant such as $\int_0^\infty \left(\iint_{\mathcal{F}} h_{t,u_0}(W)dv\right)t^{s_0-1}dt$ with necessary change of variables later; here $h_{t,u_0}(W) = \frac{1}{2}\left(\sum_{T>0}|a(T)|e^{-2\pi\sqrt{t}tr(TW)}\right)$ $\left(\sum_{\sigma\in\Gamma/\Gamma_\infty}(\text{Im}\,\sigma\langle z\rangle)^{u_0}\right)$, for $u_0 > 1$and large enough $s_0$.

**Theorem (K. Imai).** *The following statements are equivalent:*

(i) *Let* $\Lambda := \{\rho \in C|\ Re\,\rho = 1/2\} \cup \{j|\ \nu_j(-\bar{z}) = \pm\nu_j(z)$ *according as* $k$ *is even or odd*$\}$ *with conditions on* $a(T)$ *as on* $b(T)$ *as in Proposition 1 and let* $f(Z) = \sum_{T>0} a(T)e^{2\pi i tr(TZ)}$ *and* $\mathcal{M}f(W; s)$, $\Phi(s, \rho)$, $\Phi(s, j)$ *as above. Then we have* $f(-Z^{-1}) = (\det Z)^k f(Z)$ *(so that* $f$ *is a cusp form).*

(ii) $\mathcal{M}f(W; s)$ *is entire in* $s$ *and bounded in vertical strips implying the same for* $\Phi(s, \lambda)$ *for all* $\lambda \in \Lambda$ *and further* $\Phi(k - s, \lambda) = (-1)^k\Phi(s, \lambda)$ *for* $\lambda \in \Lambda$.

*Proof.*

***(i)*** $\Rightarrow$ ***(ii)*:** Let $f$ be a Siegel cusp form of degree 2 and weight $k$. Then $\mathcal{M}f(W; s) = \left(\int_0^1 + \int_1^\infty f(i\sqrt{t}W)t^s\right)\frac{dt}{t} = \int_1^\infty f((i/\sqrt{t})W)t^{-s}\frac{dt}{t} + \int_1^\infty f(i\sqrt{t}W)t^s\frac{dt}{t}$. But

$$f((i/\sqrt{t})W) = f\left((i/\sqrt{t})^t W^{-1}\begin{bmatrix}0 & -1\\ 1 & 0\end{bmatrix}\right) = f((i/\sqrt{t})^t W^{-1}) = (-1)^k t^k f(i\sqrt{t}^t W).$$

Hence $\mathcal{M}f(W; s) = \int_1^\infty f(i\sqrt{t}W)t^s\frac{dt}{t} + (-1)^k\int_1^\infty f(i\sqrt{t}^t W)t^{k-s}\frac{dt}{t}$ implying at once that $\mathcal{M}f$ is entire and bounded in vertical strips of the $s$–plane and furthermore, the functional equation $\mathcal{M}f(W, s) = (-1)^k\mathcal{M}f({}^tW, k - s)$ holds. The last relation entails immediately the assertion that $\Phi(k - s, \lambda) = (-1)^k\Phi(s, \lambda)$ for $\lambda \in \Lambda$.

***(ii)*** $\Rightarrow$ ***(i)*:** Since $v(\mathcal{F})$ is finite and $\mathcal{M}f(W; s)$ is bounded independently of $W$, it is in $L^2(H/\text{SL}(2, \mathbf{Z}))$ and has the spectral expansion $\sum_{j=0}^\infty \Phi(s, j)\nu_j + \frac{1}{4i\pi}\int_{\text{Re}\,u=1/2}\Phi(s, \rho)\text{E}_\rho(z)du$. Thus we have $\mathcal{M}f(W, s) = (-1)^k\ \mathcal{M}f({}^tW, k - s)$. Now Mellin inversion gives

$$\begin{aligned} f(i\sqrt{t}W) &= \frac{1}{2i\pi}\int_{\text{Re}\,s=c}\mathcal{M}f(W; s)t^{-s}\text{d}s = \frac{(-1)^k}{2i\pi}\int_{\text{Re}\,s=c}\mathcal{M}f\quad({}^tW, k - s)\ t^{-s}\text{d}s \\ &= \frac{(-1)^k}{2i\pi}\int_{\text{Re}\,s=k-c}\mathcal{M}f({}^tW; s)t^{s-k}\text{d}s \\ &= \frac{(-t)^{-k}}{2i\pi}\int_{\text{Re}\,s=k-c}\mathcal{M}f({}^tW; s)t^s\text{d}s = \frac{(-1)^k}{t^k}f\left(\frac{i}{\sqrt{t}}{}^tW\right) \\ &= (-1)^k t^{-k} f((i/\sqrt{t})W^{-1}), \end{aligned}$$

using

$$a\left(T\begin{bmatrix}0 & -1\\ 1 & 0\end{bmatrix}\right) = a(T)$$

and

$$ {}^tW = W^{-1}\begin{bmatrix} 0 & -1 \\ 1 & 0 \end{bmatrix}. $$

Thus $f((i\sqrt{t})W) = (\det(i\sqrt{t}W))^{-k}f(-(i\sqrt{t}W)^{-1})$, i.e. $f(-Z^{-1}) = (\det Z)^k f(Z)$ by analytic continuation, proving the claim that '$(ii) \Rightarrow (i)$'. $\diamond$

## References


[1] T. Arakawa: Dirichlet series corresponding to Siegel's modular forms, *Math. Annalen* **238** (1978) 157–173

[2] S. Boecherer: Bemerkungen ueber die Dirichletreihen von Koecher und Maass, *Math. Goettingensis*, Preprint 1987

[3] E. Hecke: Uber die Bestimmung Dirichletscher Reihen durch ihre Funktionalgleichung, *Math. Annalen* **112** (1936) 664–699

[4] K. Imai: Generalisation of Hecke's correspondence to Siegel modular forms, *American J. Math.* **102** (1980) 903–986

[5] H. Klingen: Zum Darstellungsssatz fuer Siegelsche Modulformen, *Math. Zeit.* **102** (1967) 30–43

[6] M. Koecher: Uber Dirichlet-Reihen mit Funktionalgleichung, *J. Reine Angew. Math.* **192** (1953) 1–23

[7] H. Maass: Uber eine neue Art von nichtanalytischen automorphen Funktionen und die Bestimmung Dirichletscher Reihen durch Funktionalgleichungen, *Math. Annalen* **121** (1949) 141–183

[8] H. Maass: Automorphe Funktionen von mehreren Veraenderlichen und Dirichletsche Reihen, *Hamb. Abhand.* XVI (1949) 72–100

[9] H. Maass: Siegel's Modular Forms and Dirichlet Series, *Lecture Notes in Mathematics* **216** Springer Verlag 1971

[10] S. Raghavan and J. Sengupta: On Fourier coefficients of Maass cusp forms in 3-dimensional hyperbolic space, *Proc.Indian Acad. Sci. (Math.)* **104** (1994) 77–92

[11] G. Shimura: On Eisenstein series, *Duke Math. J.* **50** (1983) 417–476

[12] A. Weil: Uber die Bestimmung Dirichletscher Reihen durch Funktionalgleichungen, *Math. Annalen* **168** (1967) 149–156

[13] R. Weissauer: Siegel modular forms and Dirichlet series, Preprint, 1986

Proc. Int. Conf. – Number Theory No. 1, 2004, pp. 97–105.

# Nonvanishing of $L$-Functions of Cusp Forms Inside the Critical Strip

A. Raghuram

Department of Mathematics, University of Iowa, 14 MacLean Hall, Iowa City, Iowa 52242, USA
e-mail: araghura@math.uiowa.edu

April 8, 2004

**Abstract.** A theorem of W. Kohnen states that the generalized Riemann hypothesis (GRH) holds on an average for holomorphic cusp forms on the upper half plane for the full modular group $SL_2(\mathbb{Z})$. In this article we prove a couple of generalizations of this theorem of Kohnen that the GRH holds on an average for holomorphic cusp forms on the upper half plane for *arbitrary* level, weight and primitive nebentypus.

The talk I gave in the international conference ANTZETA at IMSc, Chennai, India in January 2002, was on nonvanishing of certain Rankin-Selberg $L$-functions. This has in the meanwhile appeared with all details in [7]. I take this opportunity to thank the organizers of this conference for inviting me for this short and sweet meeting of which I have many pleasant memories.

**1991 Mathematics Subject Classification:** 11F66 (11F11)

## 1. Introduction

Associated to a holomorphic cusp form $f$ of weight $k$ on the upper half plane is its (completed) $L$-function $L(s, f)$ which satisfies a functional equation relating its value at $s$ to its value at $k - s$. The critical strip for such a functional equation is $(k-1)/2 < \mathrm{Re}(s) < (k+1)/2$. The generalized Riemann hypothesis states that any zero of $L(s, f)$ in the critical strip actually lies on the critical line $\mathrm{Re}(s) = k/2$. Proving this hypothesis seems to be beyond present day technology but it is not difficult to prove it *on an average*. This article has been inspired by the following theorem of Kohnen [3].

**Theorem 1.1 (Kohnen).** *Let $k$ be an even integer. Let $\{f_1, \ldots, f_{g_k}\}$ be a basis of normalized Hecke eigenforms of $S_k$ - the space of holomorphic cusp forms of weight $k$ for $SL_2(\mathbb{Z})$. Let $t_0 \in \mathbb{R}$ and let $\epsilon > 0$. Then there exists a constant $C = C(t_0, \epsilon)$ depending only on $t_0$ and $\epsilon$ such that for $k > C$ the function*

$$\sum_{i=1}^{g_k} \frac{L(s, f_i)}{\langle f_i, f_i \rangle}$$

*does not vanish for any point $s = \sigma + it_0$ with either $(k-1)/2 < \sigma < k/2 - \epsilon$ or $k/2 + \epsilon < \sigma < (k+1)/2$.*

In this article we generalize this theorem of Kohnen to cusp forms of weight $k$, level $N$ and nebentypus $\psi$. This generalization works in two directions. As above we may let the weight get large but now fixing $N$ and $\psi$ or we may fix the weight $k$ and let the level $N$ get large. Our proof is essentially Kohnen's proof of his theorem except that it needs to keep track of these additional parameters - $k$, $N$ and $\psi$. This little extra effort then pays handsome dividends - for instance as in the second generalization of keeping the weight fixed and letting the level get large. In § 2. we state the main results and also give the essence of the proof. In § 3. we set forth some preliminaries required for the proof. The proof itself is given in § 4..

There has been some work related to the results of this paper. Akbary [1] considers the question of nonvanishing of such modular $L$-functions on small discs inside the critical strip. He also counts the number of cusp forms whose $L$-functions do not vanish on such discs; however he considers cusp forms of weight $k$, *prime* level $N$ and trivial nebentypus. There is also the recent and very deep paper of Kowalski, Michel and VanderKam [5] where they compute the asymptotics of the fourth moments (with mollifiers) of $L$-functions of weight 2 and (large) prime level cusp forms. Our own results *intersects* with the results of [1] and [5]; however the techniques are very different and our hypothesis on the weight, level and nebentypus are more general. (See also the last paragraph of § 2. for some general remarks about proofs in this paper.) J.M. Sommer in an unpublished diploma thesis written under the supervision of Winfried Kohnen also gives the kernel function for cusp forms of level $N$ and arbitrary character $\chi$ and uses it to give a new proof for the holomorphic continuation and functional equation of the Hecke $L$-series in question. He also has worked out a complete proof of the convergence of the kernel series in the range $1 < \sigma < k - 1$.

## 2. Main theorems

We need some notations before we can state the results. Let $k$ and $N$ be positive integers. Since for $N = 1$ the results boil down to Kohnen's theorem we assume that $N \geq 2$. Let $\psi$ be a *primitive* character of $(\mathbb{Z}/N\mathbb{Z})^\times$ such that $\psi(-1) = (-1)^k$. Let

$$\Gamma_0(N) = \left\{ \begin{pmatrix} a & b \\ c & d \end{pmatrix} \in SL_2(\mathbb{Z}) : c \equiv 0 \pmod{N} \right\}.$$

The character $\psi$ gives a character also denoted $\psi$ of $\Gamma_0(N)$ by setting $\psi\left(\begin{pmatrix} a & b \\ c & d \end{pmatrix}\right) = \psi(d)$.

We let $S_k(N, \psi)$ denote the space of holomorphic cusp forms on the upper half plane $\mathfrak{h}$ of weight $k$, level $N$ and nebentypus $\psi$. We would like to emphasize that since $\psi$ is assumed to be primitive, every form in $S_k(N, \psi)$ is a newform. Let $r = r(k, N, \psi)$ denote the dimension of this space. For $f \in S_k(N, \psi)$ with Fourier expansion at $i\infty$ given by

$$f(z) = \sum_{n=1}^{\infty} a_n(f) e^{2\pi i n z}$$

we let $L(s, f)$ denote its completed $L$-function given by

$$L(s, f) = (2\pi)^{-s}\Gamma(s)\sum_{n=1}^{\infty}\frac{a_n(f)}{n^s}.$$

This converges absolutely and uniformly on compact subsets in some half plane and extends to an entire function of $s$.

For $f$ a primitive form in $S_k(N, \psi)$ let $f^* \in S_k(N, \bar{\psi})$ be given by $f^* = f|_k\omega_N$ where $\omega_N = \begin{pmatrix} 0 & -1 \\ N & 0 \end{pmatrix}$. Note that $f^{**} = (-1)^k f$. We also note that the first Fourier coefficient of $f^*$ is a complex number of absolute value 1 and is denoted by $w_f$. See Theorem 4.6.15 and Corollary 4.6.18 of [6]. (In fact, the number $i^k w_f$ is the root number of the cuspidal automorphic representation of $GL_2(\mathbb{A}_{\mathbb{Q}})$ associated to $f$.)

We have the functional equation for $L(s, f)$ given by

$$L(s, f) = i^k N^{-(s-k/2)} L(k - s, f^*). \tag{2.1}$$

For $f, f' \in S_k(N, \psi)$ we let $\langle f, f' \rangle$ denote the Petersson inner product of $f$ and $f'$. Let $\{f_1, \ldots, f_r\}$ denote an orthogonal basis of $S_k(N, \psi)$ consisting of *primitive forms*. (By a primitive form $f \in S_k(N, \psi)$ we mean that it is normalized such that $a_1(f) = 1$ and that it is a common eigenform for all the Hecke operators $T(n)$ with $\gcd(n, N) = 1$.) It is well known that such a basis exists. Our first theorem is the following.

**Theorem 2.2.** *Let $N \geq 2$. Let $\{f_1, \ldots, f_r\}$ denote an orthogonal basis of $S_k(N, \psi)$ consisting of primitive forms. Let $t_0 \in \mathbb{R}$ and $\epsilon > 0$. Then there exists a constant $C_1 = C_1(t_0, \epsilon)$ depending only on $t_0$ and $\epsilon$ such that for $k > C_1$ the function*

$$\sum_{i=1}^{r}\frac{L(s, f_i)}{\langle f_i, f_i \rangle\, w_{f_i}}$$

*does not vanish for any point $s = \sigma + it_0$ with $(k-1)/2 < \sigma < k/2 - \epsilon$ or $k/2 + \epsilon < \sigma < (k+1)/2$.*

As pointed out in the introduction we get another result partially generalizing Akbary's theorem [1] by now fixing the weight but letting the level $N$ get large. We state this as our second theorem.

**Theorem 2.3.** *Let $k \geq 4$. Let $\{f_1, \ldots, f_r\}$ denote an orthogonal basis of $S_k(N, \psi)$ consisting of primitive forms. Let $t_0 \in \mathbb{R}$ and $\epsilon > 0$. Then there exists a constant $C_2 = C_2(k, t_0, \epsilon)$ depending only on $k$, $t_0$ and $\epsilon$ such that for $N > C_2$ the function*

$$\sum_{i=1}^{r}\frac{L(s, f_i)}{\langle f_i, f_i \rangle\, w_{f_i}}$$

*does not vanish for any point $s = \sigma + it_0$ with $(k-1)/2 < \sigma < k/2 - \epsilon$ or $k/2 + \epsilon < \sigma < (k+1)/2$.*

**Corollary 2.4.** Consider the space $S_k(N, \psi)$ of holomorphic cusp forms of weight $k$, level $N$ and nebentypus $\psi$. Take a point $s_0$ inside the critical strip but not on the critical line. If either $k$ or $N$ is suitably large then there exists a primitive form $f \in S_k(N, \psi)$ such that $L(s_0, f) \neq 0$.

The corollary of course trivially follows from the theorems. We would like to state this corollary in a language which we think is highly suggestive and hints at a vast generalization of our theorem(s).

**Corollary 2.5.** Let $\mathfrak{f}$ be an ideal in $\mathbb{Z}$. Let $\psi$ be an idèle class character of $\mathbb{A}_{\mathbb{Q}}^{\times}/\mathbb{Q}^*$ which is trivial on positive reals. Let $\tau$ be an irreducible admissible representation of $GL_2(\mathbb{R})$. Assume that $\tau$ is a discrete series representation of weight $k$. Consider the set $S(\mathfrak{f}, \psi, \tau)$ of equivalence classes of irreducible cuspidal automorphic representations $\pi$ of $GL_2(\mathbb{A}_{\mathbb{Q}})$ subject to the conditions that the conductor $\mathfrak{f}_\pi$ is $\mathfrak{f}$, the central character $\omega_\pi$ is $\psi$ and the representation at infinity $\pi_\infty$ is $\tau$. Assume that $S(\mathfrak{f}, \psi, \tau)$ is not empty.

Now given any complex number $s_0$ such that $0 < \mathrm{Re}(s_0) < 1$ and $\mathrm{Re}(s_0) \neq 1/2$ there exists $\pi \in S(\mathfrak{f}, \psi, \tau)$ such that $L(s_0, \pi) \neq 0$ if either the norm $N(\mathfrak{f})$ or the weight $k$ is large enough.

*Proof.* Follows from Corollary 2.4 and Chapter 5 of [2]. □

The proofs of both our theorems proceed along the same lines and differ from each other only in the last step. The essence of the proof is to represent the linear form $f \mapsto L(s, f)$, i.e., to isolate a kernel function $R_s \in S_k(N, \psi)$ depending on a complex variable $s$ such that the Petersson inner product $\langle f, R_s \rangle$ is essentially $L(s, f)$. The quantity in the theorems whose nonvanishing we want to prove is essentially the first Fourier coefficient of this function $R_s$. This is the main idea in [3] where this function is given for $N = 1$ and trivial $\psi$. We construct this function more generally in equation (3.3). We mention in passing that Zagier [8] has a construction of a cusp form for $SL(2, \mathbb{Z})$ which represents (in the above sense) the symmetric square $L$-function and this is used by Kohnen and Sengupta [4] to prove a similar theorem for such $L$-functions.

## 3. Some preliminaries

To begin with we recall the Poincaré series in $S_k(N, \psi)$. (See §2.6 of [6].) Let $k \geq 3$. Define the $m^{\text{th}}$ Poincaré series to be

$$P_m(z; k, N, \psi) = \sum_{\Gamma_0(N)_\infty \backslash \Gamma_0(N)} \overline{\psi(\gamma)}\, j(\gamma, z)^{-k}\, e^{2\pi i m \gamma z}$$

for $z$ in the upper half plane, where $\Gamma_0(N)_\infty$ is the isotropy at $\infty$ inside $\Gamma_0(N)$. It is however more convenient to go modulo

$$\left\{ \begin{pmatrix} 1 & \mathbb{Z} \\ 0 & 1 \end{pmatrix} \right\} \subset \Gamma_0(N)_\infty$$

and we have the following formula

$$P_m(z; k, N, \psi) = 1/2 \sum_{\substack{(c,d)\in\mathbb{Z}^2 \\ \gcd(cN,d)=1}} \overline{\psi(d)}\ (cNz + d)^{-k}\ e^{2\pi i m\{(a_0 z + b_0)/(cNz+d)\}} \tag{3.1}$$

where for any choice of $(c, d) \in \mathbb{Z}^2$, with $\gcd(cN, d) = 1$, we fix for once and for all, a pair $a_0, b_0 \in \mathbb{Z}$ such that $a_0 d - b_0 cN = 1$.

As is well known, for $m \geq 1$, $P_m(z; k, N, \psi) \in S_k(N, \psi)$ and for any $f \in S_k(N, \psi)$ one has

$$\langle f, P_m(- \,; k, N, \psi)\rangle = \frac{a_m(f)(k-2)!}{(4\pi m)^{k-1}}. \tag{3.2}$$

Inspired by Kohnen, we now introduce a function of $z$ in the upper half plane $\mathfrak{h}$ which depends on the parameters $k$, $N$, $\psi$, and a complex number $s$ which is given by:

$$R(z; k, N, \psi, s) = \gamma_k(s) \sum_{\left(\begin{smallmatrix} a & b \\ cN & d \end{smallmatrix}\right)\in\Gamma_0(N)} \overline{\psi(d)}\ (cNz + d)^{-k}\ ((az + b)/(cNz + d))^{-s} \tag{3.3}$$

where

$$\gamma_k(s) = 1/2\ e^{\pi i s/2}\Gamma(s)\Gamma(k - s).$$

The same arguments as in [3] shows that $R(z; k, N, \psi, s)$ is convergent absolutely and uniformly on compact subsets if $1 < \mathrm{Re}(s) < k - 1$ and for $y \geq \epsilon$ and $x \leq 1/\epsilon$ for any given $\epsilon > 0$. Clearly from the definition, $R(z; k, N, \psi, s) \in S_k(N, \psi)$. Further, using Lipschitz's formula

$$\sum_{n\in\mathbb{Z}} (\tau + n)^{-s} = \frac{e^{-\pi i s/2}(2\pi)^s}{\Gamma(s)} \sum_{n\geq 1} n^{s-1} e^{2\pi i n \tau}$$

valid for $\tau$ in the upper half plane and $\sigma > 1$, one can see that

$$R(z; k, N, \psi, s) = (2\pi)^s \Gamma(k - s) \sum_{n\geq 1} n^{s-1} P_n(z; k, N, \psi). \tag{3.4}$$

Let $c_k = i^k \pi (k-2)!/2^{k-2}$. Using (3.2), (3.4), and (2.1) we have the following formula:

$$\langle f, R(\ ; k, N, \psi, \overline{s})\rangle = c_k N^{s-k/2} L(s, f^*) \tag{3.5}$$

Now let $\{f_1, \ldots, f_r\}$ be an orthogonal basis for $S_k(N, \psi)$ then (3.5) implies that

$$R(z; k, N, \psi, s) = \overline{c_k} N^{s-k/2} \sum_{i=1}^{r} \frac{\overline{L(\overline{s}, f_i^*)}}{\langle f_i, f_i\rangle} f_i.$$

Using $\langle f_i, f_i\rangle = \langle f_i^*, f_i^*\rangle$, $f^{**} = (-1)^k f$, and $\overline{c_k}(-1)^k = c_k$ we get by now going to the space $S_k(N, \overline{\psi})$ that

$$R(z; k, N, \overline{\psi}, s) = c_k N^{s-k/2} \sum_{i=1}^{r} \frac{\overline{L(\overline{s}, f_i)}}{\langle f_i, f_i\rangle} f_i^*. \tag{3.6}$$

The burden of proof lies now in computing the first Fourier coefficient of $R(z; k, N, \psi, s)$.

Before we go to the proof of the theorem we would like to record a few formulae on integral transforms and Kummer's degenerate hypergeometric function. We refer to Kohnen's paper [3] and the references therein for the proofs of these formulae.

$$\frac{1}{2\pi i}\int_{c-i\infty}^{c+i\infty} (t+\alpha)^{-\mu}(t+\beta)^{-\nu}e^{pt}\,dt = \frac{1}{\Gamma(\mu+\nu)}p^{\mu+\nu-1}e^{-\beta p}\,{}_1F_1(\mu, \mu+\nu; (\beta-\alpha)p) \tag{3.7}$$

where $\alpha, \beta, \mu, \nu, p \in \mathbb{C}$ and $\mathrm{Re}(\mu-\nu) > 0$. The function ${}_1F_1(a, b; z)$ for $a, b, z \in \mathbb{C}$ is the so-called Kummer's degenerate hypergeometric function and is defined by

$${}_1F_1(a, b; z) = 1 + \sum_{n\geq 1} \frac{a(a+1)\ldots(a+n-1)}{b(b+1)\ldots(b+n-1)}\frac{z^n}{n!}.$$

As in [3] if

$${}_1f_1(a, b; z) = \frac{\Gamma(a)\Gamma(b-a)}{\Gamma(b)}\,{}_1F_1(a, b; z)$$

then

$${}_1f_1(a, b; z) = \int_0^1 e^{zu}u^{a-1}(1-u)^{b-a-1}\,du$$

when $\mathrm{Re}(b) > \mathrm{Re}(a) > 0$. Finally, an important estimate we need follows from this last formula which is that if $\mathrm{Re}(a) > 1$, $\mathrm{Re}(b-a) > 1$ and $z \in i\mathbb{R}$ then

$$|{}_1f_1(a, b; z)| \leq 1. \tag{3.8}$$

## 4. Proof

The proofs of both Theorems 2.2 and 2.3 are essentially the same, at least till the very end and both crucially depend on the following proposition which gives the Fourier coefficients of our function $R(z; k, N, \psi, s)$.

**Proposition 4.1.** *Suppose the function* $R(z; k, N, \psi, s)$ *has*

$$R(z; k, N, \psi, s) = \sum_{n\geq 1} r_n(k, N, \psi, s)e^{2\pi inz}$$

*for its Fourier expansion. Then the Fourier coefficients are given by*

$$r_n(k,N,\psi,s)=(2\pi)^s\Gamma(k-s)n^{s-1}+i^{-k}(2\pi)^k n^{k-1}N^{s-k}/2\sum_{\substack{(a,c)\in\mathbb{Z}^2\\ ac>0\\ \gcd(a,cN)=1}}\psi(a)c^{s-k}a^{-s}\theta(a,c)$$

*where*

$$\theta(a,c)=e^{\pi is/2}e^{2\pi ina'/(cN)}\,{}_1f_1(k-s,k;2\pi in/(cNa))$$
$$+\psi(-1)e^{-\pi is/2}e^{-2\pi ina'/(cN)}\,{}_1f_1(k-s,k;-2\pi in/(cNa)).$$

*where $a'$ is the inverse of $a$ modulo $cN$.*

*Proof of Proposition* 4.1. The proof is essentially as that of Lemma 2 of [3] while keeping track of the additional parameters like $N$ and $\psi$.

We now give a sketch of the proof of this proposition. We begin with (3.3) and break up the summation as

$$R(z;k,N,\psi,s)=\sum_{\left(\begin{smallmatrix}a & b\\ cN & d\end{smallmatrix}\right)\in\Gamma_0(N)}=\sum_{c=0}+\sum_{c\neq 0}.$$

The first summation gives us using Lipschitz's formula

$$\sum_{c=0}=2\gamma_k(s)\sum_{n\in\mathbb{Z}}(z+n)^{-s}=\sum_{n\geq 1}((2\pi)^s\Gamma(k-s)n^{s-1})e^{2\pi inz}.$$

We record this contribution to the $n$-th Fourier coefficient as

$$r_n(k,N,\psi,s)=(2\pi)^s\Gamma(k-s)n^{s-1}+t(n)$$

where $t(n)$ is the contribution from the terms with $c\neq 0$ and is given by:

$$t(n)=\int_{iC}^{iC+1}\left\{\gamma_k(s)\sum_{\substack{\left(\begin{smallmatrix}a & b\\ cN & d\end{smallmatrix}\right)\in\Gamma_0(N)\\ c\neq 0}}\overline{\psi(d)}(cNz+d)^{-k}((az+b)/(cNz+d))^{-s}\right\}e^{-2\pi inz}\,dz$$

for some $C>0$. Note that since $N\geq 2$ we get that $a\neq 0$ also in the summation. For any pair $(a,c)\in\mathbb{Z}^2$ with $\gcd(a,Nc)=1$ we fix, for once and for all, a pair $(b_0,d_0)$ such that $ad_0-b_0Nc=1$. Using this we break up the summation and then interchange the integral and summation signs (which is permitted by absolute convergence) and a little modification gives us the following equation.

$$t(n)=\gamma_k(s)\sum_{\substack{(a,c)\in\mathbb{Z}^2\\ ac\neq 0\\ \gcd(a,cN)=1}}\psi(a)\int_{iC-\infty}^{iC+\infty}(cNz+d_0)^{-k}((az+b_0)/(cNz+d_0))^{-s}e^{-2\pi inz}\,dz.$$

As in [3], make the substitution $z = w - d_0/cN$ and then consider two cases when $ac > 0$ and $ac < 0$. Denoting these two pieces as $t(n) = t^+(n) + t^-(n)$, we get for $t^+(n)$

$$\gamma_k(s) \sum_{\substack{(a,c)\in\mathbb{Z}^2 \\ ac>0 \\ \gcd(a,cN)=1}} \psi(a)(cN)^{-k} e^{2\pi i n d_0/(cN)} \int_{iC-\infty}^{iC+\infty} w^{-(k-s)} \left(\frac{aw}{cN} - \frac{1}{c^2N^2}\right)^{-s} e^{-2\pi i n w}\, dw.$$

Plugging in $w = (cN/a)it$ and using the formulae for Kummer's degenerate hypergeometric functions and the definition of $\gamma_k(s)$ we get that $t^+(n)$ is given by

$$\frac{1}{2} i^{-k}(2\pi)^k n^{k-1} N^{s-k} \sum_{\substack{(a,c)\in\mathbb{Z}^2 \\ ac>0 \\ \gcd(a,cN)=1}} \psi(a) c^{s-k} a^{-s} \{e^{\pi i s/2} e^{2\pi i n a'/(cN)}\, {}_1f_1(k-s, k; 2\pi i n/(cNa))\}.$$

The terms for $ac < 0$ give us a very similar looking expression for $t^-(n)$ with a few appropriate minus signs. Put together this proves the proposition. □

*Proof of Theorems* 2.2 *and* 2.3. If $\{f_1, \ldots, f_r\}$ is an orthogonal basis of primitive forms for $S_k(N, \psi)$ then equating the first Fourier coefficients in (3.6) gives us

$$r_1(k, N, \overline{\psi}, s) = c_k N^{s-k/2} \sum_{i=1}^{r} \frac{\overline{L(\overline{s}, f_i)}}{\langle f_i, f_i\rangle} w_{f_i} = c_k N^{s-k/2} \sum_{i=1}^{r} \overline{\frac{L(\overline{s}, f_i)}{\langle f_i, f_i\rangle\, w_{f_i}}}.$$

If the right hand side vanishes then using Proposition 4.1 we get

$$-2 = i^{-k}((2\pi)^{k-s}/\Gamma(k-s)) N^{s-k} \sum_{\substack{(a,c)\in\mathbb{Z}^2 \\ ac>0 \\ \gcd(a,cN)=1}} \overline{\psi(a)} c^{s-k} a^{-s} \theta(a, c).$$

Using the functional equation it suffices to consider $s = \sigma + it$ where $(k-1)/2 < \sigma < k/2 - \epsilon$. So we put $s = k/2 - \delta - it_0$ where $\epsilon < \delta < 1/2$. Taking absolute values and using the estimate in (3.8) we get that

$$2 \le A(t_0) \left| \frac{(2\pi)^{k/2+\delta+it_0}}{\Gamma(k/2+\delta+it_0)} \right| \frac{1}{N^{k/2+\delta}} \left( \sum_{\substack{(a,c)\in\mathbb{Z}^2 \\ ac>0 \\ \gcd(a,cN)=1}} \frac{1}{a^{k/2-\delta}\, c^{k/2+\delta}} \right)$$

for a constant $A(t_0) > 0$ which depends only on $t_0$.

$$2 \le A(t_0) B \left| \frac{(2\pi)^{k/2+\delta+it_0}}{\Gamma(k/2+\delta+it_0)} \right| \frac{1}{N^{k/2+\delta}} \tag{4.1}$$

for an absolute constant $B > 0$.

For Theorem 2.2 we fix $N$ and let $k$ go to infinity and for Theorem 2.3 we fix $k$ and let $N$ go to infinity. In either case the right hand side of (4.1) goes to zero (which can be seen by using Stirling's formula in the first case and is trivial in the second case). The contradiction then proves the theorems. □

## Acknowledgements

I thank Eknath Ghate for some helpful discussions on modular forms and I thank Winfried Kohnen for some detailed comments on an earlier version.

## References

[1] A. Akbary, *Non-vanishing of modular L-functions on a disc*, Acta Arith., **92**, no. 4 (2000) 303–318
[2] S. Gelbart, *Automorphic forms on adele groups*, Annals of Mathematics Studies, no. 83, Princeton University Press (1975)
[3] W. Kohnen, *Nonvanishing of Hecke L-functions associated to cusp forms inside the critical strip*, J. Number Theory, **67**, no. 2 (1997) 182–189
[4] W. Kohnen and J. Sengupta, *Nonvanishing of symmetric square L-functions of cusp forms inside the critical strip*, Proc. Amer. Math. Soc. **128**, no. 6 (2000) 1641–1646
[5] E. Kowalski, P. Michel and J. VanderKam, *Mollification of the fourth moment of automorphic L-functions and arithmetic applications*, Inventiones Math., **142**, no. 1 (2000) 95–151
[6] T. Miyake, *Modular forms*, Springer-Verlag, Berlin, x+335 pp (1989)
[7] A. Raghuram, *Nonvanishing of certain Rankin-Selberg L-functions*, C. R. Math. Acad. Sci. Soc. R. Can. **24**, no. 2 (2002) 67–71
[8] D. Zagier, *Modular forms whose Fourier coefficients involve zeta-functions of quadratic fields*, Modular functions of one variable, VI (Proc. Second Internat. Conf., Univ. Bonn, Bonn, 1976), pp. 105–169. Lecture Notes in Math., Vol. 627, Springer, Berlin (1977)

Proc. Int. Conf. – Number Theory No. 1, 2004, pp. 107–116.

# Quadratic Reciprocity Via Theta Functions

M. Ram Murty[1]* and Allison Pacelli[2]

[1]Department of Mathematics, Queen's University, Kingston K7L 3N6, Ontario, Canada
e-mail: murty@mast.queensu.ca
[2]Department of Mathematics, Williams College, Williamstown, MA 01267
e-mail: Allison.Pacelli@williams.edu

## 1. Introduction

The celebrated law of quadratic reciprocity has had numerous proofs. Gauss, who first discovered the law, gave several proofs in his famous book, *Disquitiones Arithmeticae*. A proof, not widely known, is due to Hecke and it uses $\theta$-functions in number fields. In Chapter 8 of his classical text [H], the complete proof using $\theta$-functions of several variables is given in its full generality. Among the experts, it is widely believed that this proof is conceptually significant and leads to natural generalizations in the higher dimensional context (see for example [Kub]). From a didactic standpoint, it seems desirable to make Hecke's proof accessible to the undergraduate. This is the purpose of this paper.

In this task, we are simply not going to follow Hecke and merely specialize to the case of the rational number field. Rather, we expose the underlying idea and simplify considerably much of his derivation.

Let us recall the statement of the law of quadratic reciprocity. If $p$ and $q$ are distinct odd primes, we define the Legendre symbol $\left(\frac{p}{q}\right)$ to be $+1$ if the congruence $x^2 \equiv p \pmod{q}$ has a solution; we define it to be $-1$ if the congruence has no solution.

**Theorem 1 (The Law of Quadratic Reciprocity).**

$$\left(\frac{p}{q}\right)\left(\frac{q}{p}\right) = (-1)^{\frac{p-1}{2}\frac{q-1}{2}}. \tag{1}$$

The theorem is remarkable in many ways, the most notable being the relationship between the solvability of $x^2 \equiv p \pmod{q}$ to the solvability of $x^2 \equiv q \pmod{p}$.

Hecke's proof makes use of the classical $\theta$-function:

$$\theta(t) = \sum_{n=-\infty}^{\infty} e^{-n^2\pi t} \tag{2}$$

*Research partially supported by NSERC.

and its functional equation

$$\theta\left(\frac{1}{t}\right) = t^{1/2}\theta(t). \tag{3}$$

In fact, the central point of Hecke is that (3) implies (1), and it is this feature we will expose in this paper.

Historically, there seems to be some precedence for Hecke's proof. In his famous paper of 1860, Riemann used (3) to derive the analytic continuation and functional equation of the $\zeta$-function, defined as

$$\zeta(s) = \sum_{n=1}^{\infty} \frac{1}{n^s} \tag{4}$$

for $\Re(s) > 1$. Shortly thereafter, $\theta$-functions became more widely used in other contexts. But the idea that (3) implies (1) seems to originate with Landsberg [Lan] in 1893. By a complicated process, he derives using (3), the amazing identity

$$\frac{1}{\sqrt{p}} \sum_{r=0}^{p-1} e^{2\pi i r^2 q/p} = \frac{e^{\pi i/4}}{\sqrt{2q}} \sum_{r=0}^{2q-1} e^{\frac{-\pi i r^2 p}{2q}} \tag{5}$$

valid for any two coprime numbers $p$ and $q$. Apparently, this identity was first discovered by Schaar [Sc] by another method. Recently, the authors in [AR] give a new proof of this identity using quantum mechanics.

Experts will recognize the left hand side of (5) as a Gauss sum and that the identity immediately leads to:

**Theorem 2.** *Let b be any odd number. Then,*

$$\sum_{r=0}^{b-1} e^{2\pi i r^2/b} = \begin{cases} \sqrt{b} & \text{if } b \equiv 1 \pmod 4 \\ i\sqrt{b} & \text{if } b \equiv 3 \pmod 4 \end{cases}. \tag{6}$$

*Proof.* We set $q = 1$ in (5) and the result is immediate. □

The method of deriving Theorem 1 from Theorem 2 is well-known. However, for the sake of completeness, we review it below.

## 2. Gauss sums

For any natural number $q$, consider the group $(\mathbb{Z}/q\mathbb{Z})^*$ of coprime residue classes (mod $q$). A homomorphism

$$\chi : (\mathbb{Z}/q\mathbb{Z})^* \to \mathbb{C}^*$$

is called a **Dirichlet character**. The **Gauss sum** attached to $\chi$ and the residue class $n$ (mod $q$) is by definition

$$G(\chi, n) = \sum_{\substack{a=1 \\ (a,q)=1}}^{q} \chi(a) e^{2\pi i a n/q}. \tag{7}$$

Observe that when $(n, q) = 1$, we have

$$\chi(n)G(\chi, n) = \sum_{\substack{a=1\\(a,q)=1}}^{q} \chi(n)\chi(a)e^{2\pi ian/q} = \sum_{\substack{b=1\\(b,q)=1}}^{q} \chi(b)e^{2\pi ib/q} = G(\chi, 1), \tag{8}$$

because $\chi$ is a homomorphism and as $a$ runs through the coprime residue classes, so does $an$. In the special case that $q$ is prime and $\chi$ is the Legendre symbol, observe that:

$$\sum_{\substack{b=1\\(b,q)=1}}^{q} (1 + \chi(b))e^{2\pi ib/q} = \sum_{j=1}^{q-1} e^{2\pi ij^2/q} \tag{9}$$

as $\chi(b) = \pm 1$ according as $b$ is a square (mod $q$) or not. Since

$$1 + \sum_{b=1}^{q-1} e^{2\pi ib/q} = 0,$$

we deduce

$$\sum_{\substack{b=1\\(b,q)=1}}^{q} \chi(b)e^{2\pi ib/q} = \sum_{j=0}^{q-1} e^{2\pi ij^2/q}. \tag{10}$$

More generally, if $(a, q) = 1$, then

$$\sum_{\substack{b=1\\(b,q)=1}}^{q} (1 + \chi(b))e^{2\pi iba/q} = \sum_{j=1}^{q-1} e^{2\pi ij^2a/q} \tag{11}$$

from which we deduce

$$\sum_{\substack{b=1\\(b,q)=1}}^{q} \chi(b)e^{2\pi iba/q} = \sum_{j=0}^{q-1} e^{2\pi ij^2a/q}. \tag{12}$$

From Theorem 2, we deduce:

**Theorem 3.** *Let $q$ be prime. Then*

$$\sum_{\substack{b=1\\(b,q)=1}}^{q} \left(\frac{b}{q}\right)e^{2\pi ib/q} = \begin{cases} \sqrt{q} & \text{if } q \equiv 1 (\text{mod } 4)\\ i\sqrt{q} & \text{if } q \equiv 3 (\text{mod } 4)\end{cases} \tag{13}$$

The left hand side of the equation in Theorem 3 is a Gauss sum, and the theorem provides an explicit determination of it.

If we define for any natural number $b$,

$$S(b, a) = \sum_{j=0}^{b-1} e^{2\pi i j^2 a/b} \tag{14}$$

then

$$S(ab, 1) = S(a, b)S(b, a). \tag{15}$$

To see this, observe that

$$S(ab, 1) = \sum_{j=0}^{ab-1} e^{\frac{2\pi i j^2}{ab}} = \sum_{j_1=0}^{a-1} \sum_{j_2=0}^{b-1} e^{\frac{2\pi i (bj_1+aj_2)^2}{ab}} \tag{16}$$

since every residue class $j \pmod{ab}$ can be written as $bj_1 + aj_2$ with $0 \le j_1 \le a-1$ and $0 \le j_2 \le b-1$. Expanding the square and simplifying the identity (15) is immediate.

We are now in a position to deduce Theorem 1. From (12), we have for a quadratic character $\chi$,

$$S(q, a) = \sum_{\substack{b=1 \\ (b,q)=1}}^{q} \chi(b) e^{2\pi i ba/q} = \chi(a) \sum_{\substack{b=1 \\ (b,q)=1}}^{q} \chi(b)\chi(a) e^{2\pi i ba/q} \tag{17}$$

as $\chi^2(a) = 1$. The sum on the right hand side is

$$\chi(a) \sum_{\substack{b=1 \\ (b,q)=1}}^{q} \chi(ba) e^{2\pi i ba/q} = \chi(a) G(\chi, 1). \tag{18}$$

Therefore, by Theorem 3,

$$S(q, a) = \left(\frac{a}{q}\right) \epsilon(q)\sqrt{q}$$

where $\epsilon(q) = 1$ or $i$ according as $q \equiv 1 \pmod 4$ or $3 \pmod 4$. By (15),

$$S(pq, 1) = \left(\frac{p}{q}\right)\left(\frac{q}{p}\right) \epsilon(p)\epsilon(q)\sqrt{pq}. \tag{19}$$

By Theorem 2, the left hand side is $\epsilon(pq)\sqrt{pq}$. Thus,

$$\epsilon(pq) = \left(\frac{p}{q}\right)\left(\frac{q}{p}\right) \epsilon(p)\epsilon(q), \tag{20}$$

from which Theorem 1 is immediate.

## 3. Poisson's summation formula

It remains to prove the $\theta$-function identity (3) and deduce (5) from it. The usual tool for deriving (3) is Poisson's summation formula.

Recall Féjer's fundamental theorem [Mur, 69–72] concerning Fourier series. Let $f(x)$ be a function of a real variable which is bounded, measurable, and periodic with period 1. The Fourier coefficients of $f$ are, by definition, given by

$$c_n = \int_0^1 f(x)e^{-2\pi inx}dx, \tag{21}$$

for each $n$ in $\mathbb{Z}$. The partial sums of the Fourier series of $f$ are defined as

$$S_N(x) = \sum_{|n|\leq N} c_n e^{2\pi inx}. \tag{22}$$

Let $x_0 \in \mathbb{R}$ be such that $f(x)$ admits left and right limits there:

$$f(x_0 \pm 0) = \lim_{h\to 0^+} f(x_0 \pm h). \tag{23}$$

Then Féjer proved

$$\frac{f(x_0+0)+f(x_0-0)}{2} = \lim_{N\to\infty} \frac{S_0(x)+S_1(x)+\cdots+S_N(x)}{N+1}. \tag{24}$$

If $f(x)$ is continuous at $x_0$, and the partial sums $S_N(x_0)$ converge, then

$$f(x_0) = c_0 + \sum_{n=1}^{\infty}(c_n e^{2\pi inx} + c_{-n}e^{-2\pi inx}). \tag{25}$$

When $f(x)$ is continuous and $\sum_{n=-\infty}^{\infty} |c_n| < \infty$, then the function is represented by the absolutely convergent Fourier series:

$$f(x) = \sum_{n=-\infty}^{\infty} c_n e^{2\pi inx}. \tag{26}$$

If $F(x)$ is continuous such that

$$\int_{-\infty}^{\infty} |F(x)|dx < \infty, \tag{27}$$

then we define its Fourier transform by

$$\hat{F}(u) = \int_{-\infty}^{\infty} F(x)e^{-2\pi ixu}dx. \tag{28}$$

It is also a continuous function of $u$, and if

$$\int_{-\infty}^{\infty} |\hat{F}(u)|du < \infty, \tag{29}$$

then we have the Fourier inversion formula:

$$F(x) = \int_{-\infty}^{\infty} \hat{F}(u)e^{2\pi ixu}du. \tag{30}$$

**Lemma 1.**

$$\int_{-\infty}^{\infty} e^{-\pi x^2} dx = 1. \tag{31}$$

*Proof.* We have upon squaring the left hand side

$$\int_{-\infty}^{\infty}\int_{-\infty}^{\infty} e^{-\pi(x^2+y^2)}\, dx\, dy.$$

Switching to polar coordinates ($x = r\cos\theta$, $y = r\sin\theta$) and evaluating the Jacobian, the above integral becomes

$$\int_0^{\infty}\int_0^{2\pi} e^{-\pi r^2} r\, d\theta\, dr = 2\pi \int_0^{\infty} e^{-\pi r^2} r dr = 1, \tag{32}$$

as desired. □

**Lemma 2.** *For any $u \in \mathbb{R}$,*

$$\int_{-\infty}^{\infty} e^{-\pi(x+iu)^2} dx = 1. \tag{33}$$

*Proof.* Observe that

$$\begin{aligned} \frac{\partial}{\partial u}\int_0^{\infty} e^{-\pi(x+iu)^2} dx &= \int_{-\infty}^{\infty}\Big(\frac{\partial}{\partial u} e^{-\pi(x+iu)^2}\Big) dx \\ &= 2\pi i \int_{-\infty}^{\infty} (x+iu) e^{-\pi(x+iu)^2} dx \\ &= i \int_{-\infty}^{\infty} \frac{\partial}{\partial x} e^{-\pi(x+iu)^2} dx \\ &= \Big[ i e^{-\pi(x+iu)^2}\Big]_{x=-\infty}^{x=+\infty} = 0, \end{aligned} \tag{34}$$

so that the integral is independent of $u$. Setting $u = 0$ and using Lemma 1 gives the desired result. □

We now derive:

**Theorem 4.** *Let $F \in L^1(\mathbb{R})$. Suppose that the series $\sum_{n\in\mathbb{Z}} F(n+v)$ converges absolutely and uniformly in $v$ and that $\sum_{m\in\mathbb{Z}} |\hat{F}(m)| < \infty$. Then*

$$\sum_{n\in\mathbb{Z}} F(n+v) = \sum_{n\in\mathbb{Z}} \hat{F}(n) e^{2\pi i n v}. \tag{35}$$

*Proof.* The function

$$G(v) = \sum_{n\in\mathbb{Z}} F(n+v) \tag{36}$$

is a continuous function of $v$ and of period 1. The Fourier coefficients of $G$ are given by

$$\begin{aligned} c_m &= \int_0^1 G(v)e^{-2\pi i m v}dv \\ &= \sum_{n\in\mathbb{Z}} \int_0^1 F(n+v)e^{-2\pi i m v}dv \\ &= \sum_{n\in\mathbb{Z}} \int_n^{n+1} F(x)e^{-2\pi i m x}dx \\ &= \int_{-\infty}^{\infty} F(x)e^{-2\pi i m x}dx \\ &= \hat{F}(m). \end{aligned} \tag{37}$$

Since $\sum_{m\in\mathbb{Z}} |\hat{F}(m)| < \infty$, we can represent $G$ by its Fourier series:

$$\sum_{n\in\mathbb{Z}} F(n+v) = \sum_{n\in\mathbb{Z}} \hat{F}(n)e^{2\pi i n v}, \tag{38}$$

as desired. □

**Corollary 1 (Poisson's summation formula).** With $F$ as in Theorem 4, we have

$$\sum_{n\in\mathbb{Z}} F(n) = \sum_{n\in\mathbb{Z}} \hat{F}(n). \tag{39}$$

*Proof.* Set $v = 0$ in the Theorem. □

Now we can deduce the functional equations of the $\theta$-function.

**Theorem 5.** *For $t > 0$,*

$$\sum_{n\in\mathbb{Z}} e^{-n^2\pi/t} = t^{\frac{1}{2}} \sum_{n\in\mathbb{Z}} e^{-n^2\pi t}. \tag{40}$$

*Proof.* We apply Corollary 1 to the function $F(x) = e^{-\pi x^2 t}$. It clearly is an element of $L^1(\mathbb{R})$ and the series

$$\sum_{n\in\mathbb{Z}} e^{-\pi(n+v)^2 t}$$

converges absolutely and uniformly in $v$. Moreover,

$$\begin{aligned} \hat{F}(m) &= \int_{-\infty}^{\infty} e^{-\pi x^2 t} e^{-2\pi i x m} dx \\ &= e^{-\pi m^2/t} \int_{-\infty}^{\infty} e^{-\pi(x\sqrt{t}+im/\sqrt{t})^2} dx \\ &= t^{-1/2} e^{-\pi m^2/t}, \end{aligned} \tag{41}$$

by Lemma 2. Thus $\sum_{m\in\mathbb{Z}} |\hat{F}(m)| < \infty$, and we can apply Corollary 1 to deduce the result. □

If we define for $z \in \mathbb{C}$, with $\Im(z) > 0$,

$$\Theta(z) = \sum_{n \in \mathbb{Z}} e^{i\pi n^2 z},$$

then (40) says that

$$\Theta\Big(-\frac{1}{it}\Big) = t^{1/2}\Theta(it).$$

By analytic continuation, we deduce

$$\Theta\Big(-\frac{1}{z}\Big) = \Big(\frac{z}{i}\Big)^{1/2}\Theta(z). \tag{40$'$}$$

## 4. The Landsberg–Schaar identity

It now only remains to prove (5). In addition to Theorem 5, we will need:

**Lemma 3.** *Let $a, b \in \mathbb{Z}$. Let $f(x)$ be a differentiable function on $[a, b]$. Then*

$$\frac{1}{2}f(a) + f(a+1) + \cdots + f(b-1) + \frac{1}{2}f(b) = \int_a^b f(t)dt + O\Big(\int_a^b |f'(t)|dt\Big). \tag{42}$$

*Remark.* This is a special case of the well-known Euler-Maclaurin summation formula (see for example, [Mur, 21]).

*Proof.* We have

$$\int_n^{n+1} \Big(t - n - \frac{1}{2}\Big) f'(t)dt = \frac{f(n) + f(n+1)}{2} - \int_n^{n+1} f(t)dt. \tag{43}$$

Summing this identity over the range $a \leq n \leq b-1$ and noting that $|t - n - \frac{1}{2}| \leq 1$ for $t \in [n, n+1]$ gives the desired result. □

As a corollary, we deduce:

**Lemma 4.**

$$\sum_{t=-\infty}^{\infty} e^{-(b+tp)^2\epsilon} \sim \frac{1}{p}\sqrt{\frac{\pi}{\epsilon}} \tag{44}$$

*as $\epsilon \to 0$.*

*Proof.* We apply Lemma 3 to $f(t) = e^{-(b+tp)^2\epsilon}$ for $t \in [-N, N]$ and let $N \to \infty$, to deduce

$$\sum_{t=-\infty}^{\infty} e^{-(b+tp)^2\epsilon} = \int_{-\infty}^{\infty} e^{-(b+tp)^2\epsilon}dt + O(\epsilon). \tag{45}$$

By Lemma 1, the integral equals $\frac{\sqrt{\pi}}{p\sqrt{\epsilon}}$ from which the lemma follows. □

*Remark.* Notice that the lemma is valid for complex values of $\epsilon$ provided that $-\frac{\pi}{2} < \arg\epsilon < \frac{\pi}{2}$, that is for $\Re(\epsilon) > 0$.

Now consider

$$\theta\Big(\epsilon - \frac{2iq}{p}\Big) = \sum_{n=-\infty}^{\infty} e^{-\pi n^2(\epsilon - 2iq/p)} = \sum_{b=0}^{p-1} e^{2\pi i b^2 q/p}\Bigg(\sum_{n\equiv b (\mathrm{mod}\ p)} e^{-\pi n^2 \epsilon}\Bigg). \tag{46}$$

By Lemma 4, the inner sum on the right hand side of (46) is asymptotic to $\frac{\sqrt{\pi}}{p\sqrt{\epsilon}}$ as $\epsilon \to 0^+$. Therefore, as $\epsilon \to 0^+$,

$$\theta\Big(\epsilon - \frac{2iq}{p}\Big) \sim \frac{\sqrt{\pi}\, S(p,q)}{p\sqrt{\epsilon}} \tag{47}$$

in the notation of (14). By (40$'$),

$$\theta\Bigg(\frac{1}{\epsilon - \frac{2iq}{p}}\Bigg) = \Big(\epsilon - \frac{2iq}{p}\Big)^{1/2}\theta\Big(\epsilon - \frac{2iq}{p}\Big)$$

so that

$$\frac{\sqrt{\pi}\, S(p,q)}{p\sqrt{\epsilon}} \sim \Big(\epsilon - \frac{2iq}{p}\Big)^{-1/2}\theta\Bigg(\frac{1}{\epsilon - \frac{2iq}{p}}\Bigg) \tag{48}$$

as $\epsilon \to 0^+$. If we define $\tau$ by the equation

$$\frac{1}{\epsilon - 2iq/p} = \frac{\epsilon + 2iq/p}{\epsilon^2 + 4q^2/p^2} = \tau + \frac{ip}{2q}, \tag{49}$$

then $\tau \to 0$ as $\epsilon \to 0^+$. Moreover, $\Re(\tau) > 0$, for $\epsilon > 0$. We have

$$\begin{aligned}\theta\Bigg(\frac{1}{\epsilon - \frac{2iq}{p}}\Bigg) &= \theta\Big(\tau + \frac{ip}{2q}\Big)\\ &= \sum_{n=-\infty}^{\infty} e^{-\pi n^2(\tau + ip/2q)}\\ &= \sum_{b=0}^{4q-1} e^{-2\pi i b^2 p/4q}\Bigg(\sum_{n\equiv b\ (\mathrm{mod}\ 4q)} e^{-\pi n^2 \tau}\Bigg).\end{aligned} \tag{50}$$

Notice that as $\tau \to 0$, the inner sum on the right hand side of (50) can be treated as in (46) and we deduce

$$\theta\Big(\frac{1}{\epsilon - 2iq/p}\Big) \sim \frac{S(4q,-p)\sqrt{\pi}}{4q\sqrt{\tau}} \tag{51}$$

as $\epsilon \to 0^+$. From (49),

$$\frac{\tau}{\epsilon} = -\frac{ip}{2q(\epsilon - 2iq/p)}$$

so that from (48), we obtain

$$\frac{S(p,q)}{p}\sqrt{\frac{-ip}{2q}} = \frac{S(4q,-p)}{4q}. \tag{52}$$

Now $\sqrt{-i} = e^{-i\pi/4}$, the determination being ascertained in several ways. (For example, take $p = q = 1$ in (52) to determine it.) Upon noting that

$$\sum_{b=o}^{4q-1} e^{\frac{-2\pi i p b^2}{4q}} = 2\sum_{b=0}^{2q-1} e^{\frac{-\pi i b^2 p}{2q}}$$

we immediately obtain (5). This completes the proof of the Landsberg – Schaar identity.

## 5. Concluding remarks

It is possible to give an elementary derivation of the $\theta$-function transformation law by Polya [Po] (see also [Be]).

The proof given above appears with very few details in two places. The first is [DM] and the second is [Be]. It is also mentioned in several places such as [Te], [BEW], and [BE]. We hope that our exposition makes this idea more widely known and supplies the relevant details of the proof which are omitted in the literature.

We also refer the reader to the website: http://www.rzuser.uni-heidelberg.de/~hb3/rchrono.html for an extensive history and references of the proofs of the law of quadratic reciprocity.

## References

[AR] V. Armitage and A. Rogers, Gauss sums and quantum mechanics, *J. Phys. A.* **33** no. 34 (2000) 5993–6002

[Be] R. Bellman, *A Brief Introduction to Theta Functions*, (Rinehart & Winston) (1961)

[BE] B. Berndt and R. Evans, Determination of Gauss sums, *Bulletin of the Amer. Math. Society.* **5** no. 2 (1981) 107–129

[BEW] B. Berndt, R. Evans and K. S. Williams, Gauss and Jacobi Sums, *Canadian Mathematical Society Series of Monographs and Advanced Texts* **21**, (Wiley Interscience, New York) (1998)

[DM] H. Dym and H. P. McKean Jr., Fourier Series and Integrals, *Probability and Mathematical Statistics*, (Academic Press) no. 14 (1972)

[H] E. Hecke, *Lectures on the Theory of Algebraic Numbers*, Springer-Verlag, (1981)

[Kub] T. Kubota, On Automorphic Functions and the Reciprocity Law in a Number Field, *Lectures in Mathematics*, 2 (1969)

[Lan] G. Landsberg, Zur Theorie der Gausschen Summen und der linearen Transformation der Thetafunctionen, *J. Reine Angew Math.* **111** (1893) 234–253

[Mur] R. Murty, Problems in Analytic Number Theory, *Graduate Texts in Mathematics*, Springer-Verlag, (2001)

[Po] G. Polya, Elementarer Beweis einer Thetaformel, Sitz. der *Phys-Math Klasse*, (Berlin) (1927) 158–161 (see also Collected Papers, vol. 1 303–306, MIT Press, 1974)

[Sc] M. Schaar, Mémoire sur la théorie des résidus quadratiques, *Acad. Roy. Sci. Lettres Beaux Arts Belgique* **24** (1850) 14p

[Te] A. Terras, *Harmonic Analysis on Symmetric Spaces and Applications*, vol. 1, Springer-Verlag, (New York) (1985)

Proc. Int. Conf. – Number Theory No. 1, 2004, pp. 117–120.

# On an Arithmetical Inequality, III

S. Srinivasan

School of Mathematics, TIFR Centre, IISc Campus, Bangalore 560 012, India
e-mail: ssvasan@math.tifrbng.res.in

**Abstract.** In earlier parts it was seen that a well known Conjecture of Alladi-Erdös-Vaaler follows from lower bounds for certain discrete sums. Here we show (in a general setting) that there are related integrals, corresponding to these discrete sums, having magnitudes of the same order (up to absolute constant multiples).

## 1. Introduction

Let $\nu = 0$ or 1, and $0 < \alpha < 1$. Set $U = \{1, \ldots, k\}$, and for given reals $x_j, \xi_j, (j \in U)$ with $\xi_j > 0$, define $x_S = \Sigma_{j \in S} x_j$ and $\xi_S = \Pi_{j \in S} \xi_j$, where $S \subseteq U$. Suppose that $x_U = \nu$, and that $x_j$'s are positive, if $\nu = 1$.

Let $h$ be a non-negative submultiplicative function defined on natural numbers satisfying, $h(p) \leq c$, for primes $p$ and some fixed $c > 0$. Let $0 < \alpha < 1$. With squarefree integers $n$, we denote by $\mathbf{K(c}, \alpha)$ the following statement: The partial sum of values $h(d)$ over divisors $d$ of $n$ not exceeding $n^\alpha$ is more than a constant $\kappa$ times the complete sum of values $h(d)$ over all divisors $d$ of $n$, with a positive $\kappa = \kappa(c, \alpha)$.

In 1987, Alladi, Erdös and Vaaler conjectured in [1] that $\mathbf{K(c}, \alpha)$ is true for $\alpha = c/(c+1)$, while $c$ is the reciprocal of an integer. For details of later work, see [4]. Here we recall that they also conjectured later that perhaps $\kappa$ is an absolute constant (as $c \leq 1$). And we have, in view of [3], that the conjecture holds true, if established for multiplicative functions $h$ satisfying $h(p) = c$ (for all primes $p$). This means that we need lower bounds for discrete sums of the form, in the above notation (with $\nu = 1$):

$$\sum_{x_S \leq \alpha, S \subseteq U} \alpha^{|S|}(1-\alpha)^{k-|S|}.$$

In the present paper we associate, with (more general) sums of this form, integrals having same numerical order (cf., Section 2). The interesting part is that all these (more general) integrals involve a (unique) function $R$ arising from a certain non-linear differential equation (E) below.

## 2. Main result

For **any** given (real) $\xi \neq -1$, define the function $R$ (over reals $t$) by $2R'(t) = r^{-2}(t)$ and $R(0) = 0$, where $r^2(t) = (1+\xi)^2 - 4\xi \sin^2(t/2)$; and let $R_j$ denote the function

$R$ for $\xi = \xi_j$. Assume that $\gamma$ is chosen subject to $\gamma < x_S$, for all $S \subseteq U$, and set $\delta_1 = \frac{1}{2}(\nu\alpha - \gamma)$; $\delta_2 = \frac{1}{2}(\nu - \nu\alpha - \gamma)$.

Introduce the sums

$$\Omega_\nu = \sum_{S \subseteq U}^{(\nu)} \xi_S, \tag{1}$$

where $(\nu)$ denotes the condition that $x_S \leq \nu\alpha$. Then we obtain the following

**Theorem.** *We have*

$$\Omega_\nu \asymp J_\nu, \quad (\nu = 0, 1) \tag{2}$$

*where the implied order-constants are absolute, and*

$$J_\nu = \int_0^\infty \frac{\sin \delta_1 t}{t} \frac{\cos\left(\delta_2 t + \sum_j \lambda_j R_j(x_j t)\right)}{(R_1'(x_1 t) \ldots R_k'(x_k t))^{1/2}} dt,$$

*with* $\lambda_j = \frac{1}{2}(\xi_j{}^2 - 1)$.

## 3. A lemma

For a proof of the Theorem, we require the following Lemma. It has been that, when the details of its proof were not completed, the following differential equation (E) was also (independently) more extensively studied in [2]. (In the sequel $'$ denotes differentiation with respect to $t$.)

**Lemma.** *For* **any** *given (real)* $\xi \neq -1$, *define (real) differentiable functions* $r$, $s$ *of* $t$ *by*

$$\xi e^{it} + 1 = r e^{is}, \quad r \geq 0. \tag{3}$$

*Then we have*

$$r^2(t) = (1 + \xi)^2 - 4\xi \sin^2(t/2); \quad 2s'(t) = 1 + c_0 r^{-2}(t), \tag{4}$$

*where the constant* $c_0$ *is determined by* $s'(0) = \frac{\xi}{\xi+1}$.

*Proof.* First we rewrite (3) as

$$1 + \xi \cos t = r \cos s; \quad \xi \sin t = r \sin s. \tag{3'}$$

This immediately gives the expression for $r^2(t)$ as in (4). Further we get, by differentiation, from (3$'$),

$$r' \cos s = r(s' - 1) \sin s; \quad 2rr' = (r^2)' = -2r \sin s.$$

This yields

$$r' = -\sin s; \quad r = (1 - s')^{-1} \cos s, \tag{5}$$

leading to the differential equation

$$((1-s')^{-1}\cos s)' = -\sin s, \tag{E}$$

and the value of $s'(0)$ as given in the Lemma (by taking $t = 0$). Next, on simplification, we obtain

$$s''[(1-s')^{-1} + (2s'-1)^{-1}] = s' \tan s.$$

This gives, on integration,

$$(2s'-1)^{1/2}(1-s')^{-1} = c_1(\cos s)^{-1}$$

with a constant $c_1$, which leads to the second part of (4) with $c_0 = c_1^2$, by using (5). Since (4) implies that $s$ is twice differentiable, the proof of the Lemma is completed.

*Note.* Observe here that $s(0)$ can be (suitably) fixed anyway and so, we shall assume from here onwards that $s(0) = 0$, if $\xi > 0$.

## 4. Proof of the theorem

In the sequel we continue with notation introduced above. Set

$$\Omega_\nu^* = \sum_{S\subseteq U} \epsilon_S \xi_S; \quad 2\epsilon_S = (1 - \operatorname{sgn}(x_S - \nu\alpha)). \tag{6}$$

Obviously $\Omega_\nu \le 2\Omega_\nu^* \le 2\Omega_\nu$. Now, we have

$$a\operatorname{sgn}(\beta) = \int_0^\infty \frac{\sin \beta t}{t} dt = \int_0^\infty \frac{\operatorname{Im}(e^{i\beta t})}{t} dt,$$

with the constant $a > 0$. Thus

$$2a\Omega_\nu^* = \int_0^\infty \frac{\operatorname{Im} F(t)}{t} dt,$$

where

$$F(t) = (e^{-i\gamma t} - e^{-i\nu\alpha t}) \prod_{j=1}^{k} (\xi_j e^{i x_j t} + 1).$$

This leads, by a direct use of Lemma, to

$$a\Omega_\nu^* = \int_0^\infty \frac{\sin \delta_1 t}{t} \frac{\cos\left(\delta_2 t + \sum_j \lambda_j R_j(x_j t)\right)}{(R_1'(x_1 t)\ldots R_k'(x_k t))^{1/2}} dt = J_\nu, \tag{7}$$

say, with $\delta_1 = \frac{1}{2}(\nu\alpha - \gamma)$; $\delta_2 = \frac{1}{2}(\nu - \nu\alpha - \gamma)$. Thus the proof of the Theorem is completed.

## 5. Corollaries

We shall now take all $\xi_j$'s as equal to $\xi(>0)$, so that each $\lambda_j$ is equal to $\frac{1}{2}c_0$. Note that we can assume $\gamma = 0$, if $\nu = 1$, because then all $x_j$'s are positive. The following corollaries are immediately deduced from the Theorem. The first one is (potentially) applicable in the context of the Alladi-Erdos-Vaaler conjecture.

**Corollary 1.** Let $\nu = 1, \xi_j = \xi > 0$, for $1 \le j \le k$, and $(1+\xi)(1-\alpha) = 1$. Then

$$\sum_{x_S \le \alpha, S \subseteq U} \alpha^{|S|}(1-\alpha)^{k-|S|} \asymp \int_0^\infty \frac{\sin(\frac{\alpha}{2}t)\cos(\Psi(t))}{t\{R'_\alpha(x_1 t)\dots R'_\alpha(x_k t)\}^{1/2}} dt,$$

where

$$\Psi(t) = \frac{(1-\alpha)t}{2} - \mu \sum_{j=1}^{k} R_\alpha(x_j t),$$

with $\mu = \left(\frac{1}{2} - \alpha\right)$ and

$$R_\alpha(u) = (1+\xi)^{-2} R(u).$$

**Corollary 2.** Let $\nu = 0, \xi_j = \xi > 0$, for $1 \le j \le k$. Then

$$\sum_{x_S \le 0, S \subseteq U} \xi^{|S|} \asymp \int_0^\infty \frac{\sin \delta t}{t} \frac{\cos(\delta t + c_0 \sum_j R(x_j t))}{\{R'(x_1 t)\dots R'(x_k t)\}^{1/2}} dt,$$

where $\delta = -\frac{1}{2}\gamma$.

*Remark.* It is implicit in [4], that the (qualitative) conclusion holds even on assuming a weaker form of (1), obtained by introducing an arbitrarily small positive constant factor on the right side, there.

## References

[1] K. Alladi, P. Erdös, and J. D. Vaaler, "Multiplicative functions and small divisors", *Progress in Math.* **70** (1987) 1–13

[2] P. S. Datti and B. R. Nagaraj, "On Certain Nonlinear Differential Equations from Number Theory" (to appear)

[3] S. Srinivasan, "On an Arithmetical Inequality", *Glasgow Math. J.* **36** (1994) 81–86

[4] S. Srinivasan, "On an Arithmetical Inequality, II", *Contemp. Math.* **210** (1997) 299–301

Proc. Int. Conf. – Number Theory No. 1, 2004, pp. 121–133.

# Some Problems and Results in Arithmetic Functions

M. V. Subbarao*

Department of Mathematical Sciences, University of Alberta Edmonton, Alberta Canada T6G 2G1
e-mail: m.v.subbarao@ualberta.ca

## 1. Introduction

In this paper, I shall confine myself to some of the problems on which I worked individually or in collaboration with others. First, I consider a long neglected problem of S.S. Pillai. Since the original problem of Pillai is easily answered as 'no' (see below), a slightly amended version of the same is posed and solved and some open problems that arise in that connection are listed – of which three are due to P. Erdös. Next, I review some of the still open famous problems and conjectures on the representation of rationals as sum of unit fractions, associated with the names of Erdös, Straus, Sierpinski and Schinzel and mention my joint work with Straus as an update on them. I also mention two conjectures made jointly by Straus and myself in 1977 with regard to the divisor function and Erdös' own conjecture in a different direction regarding the same function, and give some applications. Next we review and update the still open famous conjectures of Carmichael and Sierpinski and their extensions and analogues of the same due to L.W. Yip and myself. The final section is devoted to stating some new results and conjectures relating to a new partition theoretic function which is attracting considerable attention.

The words of P. Erdös and R.L. Graham [6] used in the introduction to their well known book of problems very well apply to the problems I mention in this paper. They wrote: "The difficulty and importance of these problems are very variable – some are only probably exercises while others are very difficult or even hopeless and may have important consequences or their solution may lead to important advances."

## 2. Pillai primes

In 1930, Pillai [15] posed the following

2.1. *Question.* Is it true that every prime divisor of $f(n) = n! + 1$ is of the form $1 \pmod{n}$?

*Supported in part by an NSERC Grant

Note that this is true for all $1 \leq n \leq 7$, but false for $n = 8$. This is because $8! + 1$ is divisible by each of the primes 61 and 661, neither of which is $\equiv 1 (\text{mod } 8)$. S. Chowla [15] earlier produced two other exceptions: $14! + 1$ and $18! + 1$, each of which is divisible by the prime 23. Note that $23 \not\equiv 1 (\text{mod } 14)$, $\not\equiv 1 (\text{mod } 18)$.

**2.2. Definition.** A prime number $p$ will be called a *Pillai prime* if there exists a natural number $n$ such that $n! + 1 \equiv 0 (\text{mod } p)$, $p \not\equiv 1 (\text{mod } n)$. The first fifteen Pillai primes are 23, 29, 59, 61, 67, 71, 79, 83, 109, 137, 139, 149, 193, 227, 233. P. Erdös and independently Subbarao proved (see [9])

**2.3. Theorem.** *The number of Pillai Primes is infinite.*

Consider positive integers $n$ such that $n! + 1 \equiv 0 (\text{mod } p)$, where $p$ is a prime $\not\equiv 1 (\text{mod } n)$. Call such integers EHS (Erdös - Hardy - Subbarao) numbers. Erdös, Hardy and I spent considerable time studying such numbers. It is easy to prove [9] that they are infinite in number. We make the

2.4. *Conjecture.* Almost all numbers are EHS numbers.

Equivalently

2.5. *Conjecture.* The density of integers $n$ such that all the prime divisors of $n! + 1$ are $\equiv 1 (\text{mod } n)$ is zero.

I believe that

2.6. *Conjecture.* The number of non EHS numbers is finite.

Perhaps, this is as difficult to settle as the question of infinititude of Mersenne or Fermat primes.

Some other open problems that we raised in [9] in this connection are the following.

**2.7.** If $\pi(P, x)$ denotes the number of Pillai primes $\leq x$, and $\pi(x)$ the number of primes $\leq x$, does $\lim\limits_{x \to \infty} \frac{\pi(P,x)}{\pi(x)}$ exist?

Numerical evidence suggests that, for large $x$, we have

$$0.4 < \frac{\pi(P, x)}{\pi(x)} < 0.6$$

**2.8.** If $g(p)$ denotes the number of integers $n <$ prime $p$ for which $n! + 1 \equiv 0 (\text{mod } p)$, is $\overline{\lim}\, g(p) = \infty$? Perhaps, $g(p) \to \infty$ for almost all $p$.

(Erdös) Perhaps, the density of primes $p$ for which $g(p) = k$ exists for every $k$, and if $e(k)$ denotes this density, $\sum\limits_{k=1}^{\infty} e(k) = 1$.

**2.9.** Are there infinitely many primes $p$ for which $p - 1$ is an EHS number?

**2.10.** (P. Erdös). If $A(x) =$ number of composite numbers $u < x$ for which there is $n$ with $n! + 1 \equiv 0 (\text{mod } u)$, is $A(x) = o(x^{\varepsilon})$? Three such examples of the numbers $u$ are 25, 121, 721.

**2.11.** (P. Erdös). Are there many primes $p$ for which $n!(\text{mod } p)$ has $p-2$ nonzero values, $p$ being a prime. One example is $p = 5$.

**2.12.** Given prime $p$, let $f(p)$ denote the smallest integer for which $\big(f(p)\big)! + 1 \equiv 0 \,(\text{mod } p)$. We believe that $f(p) = p-1$ for infinitely many primes $p$. Probably, the number of such $p \le x$ is $o(x/\log x)$.

Erdös believes that for almost all $p$, $f(p)/p$ becomes small tending to 0 as $p \to \infty$.

## 3. Simple fractions: the Erdös-Straus, Sierpinski, and Schinzel conjectures

In 1948, Erdös and Straus made the famous, still not completely solved

3.1. *Conjecture.* For all $n > 1$,

$$\frac{4}{n} = \frac{1}{x} + \frac{1}{y} + \frac{1}{z}$$

has positive integral solutions.

This is known to hold for all $n < 10^8$ and for $n$ which belongs to any one of several arithmetic progressions. See, Mordell [13], among numerous others.

3.2. *Sierpinski's Conjecture.*

$$\frac{5}{n} = \frac{1}{x} + \frac{1}{y} + \frac{1}{z}$$

is solvable in natural numbers $x, y, z$ for all $n > 2$.

This is known to hold for all $n$ satisfying $3 \le n < 1,057,438,801$ (Stewart [19], Stewart and Webb [20]).

More generally, we have the

3.3. *Conjecture (A. Schinzel and W. Sierpinski, 1956).* The equation

$$\frac{a}{n} = \frac{1}{x} + \frac{1}{y} + \frac{1}{z}$$

is always solvable for all $n > n_0(a)$.

In 1970, R.C. Vaughan [26] proved that for any given $a > 0$, the number of $n \le x$ for which $a/n$ does not have the desired representation is $< x \exp\{-c(\log x)^{2/3}\}$ for some $c = c(a) > 3$ depending on $a$.

Thus all the three conjectures (3.1), (3.2) and (3.3) are true for almost all values of $n$.

Curiously however, we do not know explicitly any value of $a$ for which Conjecture 3.3 holds for all sufficiently large $n$.

A more tractable conjecture has been made by Schinzel in 1956, namely

3.4. *Schinzel's Conjecture.* For every integer $a$, there is a $n(a)$ such that for all $n > n(a)$, $a/n$ can be written in the form

$$\frac{a}{n} = \frac{1}{x} \pm \frac{1}{y} \pm \frac{1}{z}$$

with $x, y, z$ positive integers.

The best result known at present is that Schinzel's conjecture holds for all $a \leq 40$. This is due to Straus and Subbarao [21].

3.5. *Remark.* We might ask: given $a$, what bound can be given to $k = k(a)$ so that the equation

$$\frac{a}{n} = \frac{1}{x_1} + \frac{1}{x_2} + \cdots + \frac{1}{x_k}$$

has solutions in not necessarily positive integers $x_1, \ldots, x_k$? We have

**3.6. Theorem (Straus-Subbarao [21], William Webb [27]).** $k < c \log a$, *where* $c$ *is an absolute constant which can be explicitely given.*

**3.7. Problem** Obtain an estimate for the value of $k$ if in the above, all the integers $x_1, \ldots, x_k$ are restricted to be positive.

No non-trivial estimate for $k$ is known in this case.

## 4. The divisor function: Straus-Subbarao conjectures (1977), [21]

Let $d(n)$ denote the number of divisors of $n$.

4.1. *Conjecture.*

$$\lim_{n\to\infty} \sup_{s>0} \frac{d(n+s)}{s+1} = \infty.$$

Clearly, $\sup_s \frac{d(n+s)}{s+1}$ is attained for some $s = o(n^\varepsilon)$. What the conjecture says is, essentially, that every sufficiently large integer $n$ has a successor $n'$ with many more divisors than the distance $|n' - n|$.

As an application of this conjecture we prove

**4.2. Theorem** *Conjecture 4.1 implies the truth of the Schinzel-Sierpinski Conjecture 3.3.*

**Proof** (an outline): Write $n = am - r, 0 < r < a$ and

$$\frac{a}{n} = \frac{1}{m+s} + \frac{sa+r}{(m+s)n}, \qquad s = 0, 1, 2, \ldots \tag{4.3}$$

Conjecture 4.1 implies that

$$d(m+s) > a(s+1) > \varphi(as+r)$$

for some $s \geq 0$ when $n$- and hence $m$- is sufficiently large. Hence it follows that

**(4.4)** $m+s$ has two divisors $d_1, d_2$ – positive or negative – with $d_1 \equiv -d_2 (\mathrm{mod}\ as+r)$.

We now apply to the number $(sa+r)/(m+s)$ the following well known result (see, for example, Stewart and Webb [20], Straus and Subbarao [21, Lemma 2.1]):

**(4.5)** A rational number $A/B$ can be expressed in the form $1/x \pm 1/y$ ($x, y$ positive integers) if and only if there are distinct divisors, (positive or negative) of $B$ such that they are congruent modulo $A$.

In view of the results (4.4) and (4.5), it follows that in (4.3) $(sa+r)/(m+s)n$ has the representation $1/x \pm 1/y$ in positive integers $x, y$, thus proving Theorem 4.2.

For Schinzel's equation in 3.4, we need the weaker

4.6. *Conjecture.*

$$\lim_{n\to\infty} \sup_{s\geq 0} \frac{d(n\pm s)}{s+1} = \infty.$$

*Remark.* In order to prove the Schinzel-Sierpinski Conjecture 2.3 for specific sets $A$ of numerators $a$ we need the weaker form.

4.7. *Conjecture.*

$$\lim_{n\to\infty} \sup_{s\geq 0} \frac{d(n+s)}{\varphi(as+r)} \geq 1$$

for all $0 < r < a$, $(r, s) = 1$ and all $a \in A$.

The Schinzel Conjecture 2.4 for a specific set $A$ of $a$ follows from the much weaker form of 4.7:

4.8. *Conjecture.*

$$\lim_{n\to\infty} \sup_{s\geq 0} \frac{d(n\pm a)}{\varphi(as\pm r)} > \frac{1}{2}$$

for all $0 < r < a/2$, $(r, s) = 1$ and all $a \in A$.

When these conjectures were mentioned to P. Erdös, he conjectured a strong negation of Conjecture 4.1, namely

4.9. *Conjecture.* (Erdös, 1977). There exist arbitrarily large $n$ for which

$$d(n+s)/(s+1) \leq 2 \quad \text{for all} \quad s = 0, 1, 2, \ldots .$$

4.10. *Remark.* Erdös Conjecture implies that there exist infinitely many primes $p$ for which we have $p+1$ is twice a prime, $p+2$ a prime or thrice a prime or nine times a prime or thrice a prime squared etc.

Needless to say that all these conjectures 4.1, 4.6, 4.7, 4.8 and 4.9 are extremely deep and perhaps hopeless. I do not know if sieve methods can settle 4.9. Erdös suggested to me and Straus that we keep both our Conjecture 4.1 and his Conjecture 4.10 (though they contradict each other) with the remark that probably neither of them could be settled in our lifetimes, if at all.

4.11. *Remark.* In connection with the above conjecture, the following problem would be of great interest in itself.

**4.12. Problem (Straus-Subbarao, 1977).** For what functions $f(s)$ do we have

$$d(n+s) \leq f(s), \quad s = 0, \pm 1, \pm 2, \ldots$$

and infinitely many $n$?

4.13. *Remark.* Erdös remarked to us that sieve methods seem to suggest the existence of such functions.

Allied to the above problem, Erdös raised the following (see [6], p. 62).

**4.14. Problem (P. Erdös [6], 1980).** Is there a constant $c > 0$ so that for infinitely many $n$, we have

$$d(n+i) < ci \quad \text{for all} \quad i \geq 1.$$

Erdös ([6], p. 62) states that this inequality would help in deciding – among other things – the irrationality of $\sum_n d(n)/a_1 \cdots a_n$, for a given sequence $a_1, a_2, \cdots$, but he emphasized that "if true, it will be infinitely hard to prove". However, the above sum is proved by Erdös-Straus (see [6], p. 62) to be irrational whenever $a_n \geq a_{n-1}$ and $a_n \to \infty$ with $n$.

## 5. An extended Sierpinski conjecture

The presence of so many extensions and analogues of the Euler totient and other classical arithmetic functions would inevitably lead one to consider extensions and analogues of some of the existing celebrated problems associated with famous names.

In this and the next section, we briefly mention some of them. Sierpinski's Conjecture [18] is:

**5.1.** For any given integer $n > 1$, there are infinitely many $m$ for which the equation $\varphi(x) = m$ has exactly $n$ solutions.

Schinzel [17] showed that his hypothesis $H$ implies the truth of this conjecture.

For the convenience of the reader we state below one version of Hypothesis $H$.

**Hypothesis H.** Let $f_1(x), \ldots, f_s(x), g_1(x), \ldots, g_t(x)$ be irreducible integer valued polynomials of positive degree with positive leading coefficients. If there does not exist any integer $> 1$ dividing the product $f_1(x)f_2(x)\ldots f_s(x)$ for every $x \in \mathbb{N}$, and if

$$g_j(x) \not\equiv f_i(x) \quad \text{for all} \quad i \leq s, \ j \leq t,$$

then there exist infinitely many positive integers $x$ such that the numbers $f_1(x), \ldots, f_s(x)$ are primes, and the numbers $g_1(x), \ldots, g_t(x)$ are composite.

In 1991, L.W. Yip and I [25] extended 5.1 to Schemmel's totient $\Phi_k(n)$ defined, for $k = 1, 2, \ldots$, by $\Phi_k(1) = 1$, $\Phi_k(n) = 0$ if $n$ contains a prime factor $\leq k$, and if all the prime factors of $n$ are $> k$, then

$$\Phi_k(n) = \prod_{p^a \| n} p^{a-1}(p - k)$$

where $p^a \| n$ means $p^a | n$ and $p^{a+1} \nmid n$.

5.2. *Conjecture.* For any given natural number $k \geq 1$ and any given $n > 1$, there exist infinitely many $m$ such that the equation $\Phi_k(x) = m$ has exactly $n$ solutions.

We proved the special case of this [25]:

**5.3. Theorem.** *Under Hypothesis H, the Conjecture 5.2 holds for all integers $k$ of the form $k = p^a - 2$, where $p$ is any odd prime and $a \in \mathbb{N}$.*

We believe that the restriction on $k$ can be removed.

5.4. *Remark.* We excluded in 5.2 the case $n = 1$, which is the famous still unsolved Carmichael conjecture (which will be discussed in the next section).

5.5. *Remark.* Erdös [5] proved more than twenty five years ago: that for any given $k \geq 1$, if there is an integer $n$ for which $\varphi(m) = n$ has precisely $k$ solutions, then there are infinitely many such integers.

We believe that a similar result holds for $\Phi_k$ for all $k \geq 1$ and that Erdös' proof, with suitable changes works for this case also; but we have not carried out all the details of the proof.

## 6. On Carmichael conjecture and analogues

In the previous section, we already defined the Schemmel totient. Note that $\Phi_1$ is the Euler totient $\varphi$.

Let $N_k(m)$ denote the number of solutions of the equation $\Phi_k(x) = m$. Carmichael conjectured in 1927 [2] that $N_1(m) \neq 1$ for any $m$. This has been investigated by several authors (Erdös [5], Pomerance [16], Klee [11], Schinzel [17] and others). P. Masai and A. Valette [12] showed that there is no $x < 10^{10000}$ for which $N(\varphi(x)) = 1$. Pomerance [16] made this conjecture depend on other very deep conjectures, all of which are hopeless and open. Here we make the

6.1. *Conjecture (Subbarao-Yip [24], 1991).* $N_2(m) \neq 1$ for any $m$.

Numerical evidence is very strong. We showed in [24] that there is no $m$ for which there is an $x$ so that $\Phi_2(x) = m$ has only one solution for any $x < 10^{120000}$. Our conjecture follows if we can prove another conjecture, namely

6.2. *Conjecture.* Let $q_1 = 3$ and for each $n \geq 1$, $q_{n+1}$ is the smallest prime $> q_n$ for which $(q_{n+1} - 2) | (q_1 q_2 \ldots q_n)$. Then the sequence $\{q_n\}$ is infinite.

6.3. *Remark.* The first few terms of $\{q_n\}$ are $3, 5, 7, 17, \ldots\ q_{10000} = 4873801$. Erdös agreed with us that there appears no reason why the sequence $\{q_n\}$ should terminate.

We state without proof that if an $m$ exists for which $\Phi_2(x) = m$ has a unique solution, then $x$ is divisible by $q_n^2$ for $n = 1, 2, \ldots$, so that if $\{q_n\}$ is infinite, there is no such $x$ and hence no such $m$, and hence 6.2 implies 6.1.

As in Pomerance [16], we can find other conjectures from which 6.1 follows. For example, $N_2(\Phi_2(x)) = 1$ if there is a natural number $x$ such that for any prime $p$ for which $(p-2)|\Phi_2(x)$, we also have $p^2|x$. But such an $x$ cannot exist if the following conjecture of Subbarao-Yip [24] holds:

6.4. *Conjecture.* Let $p_i$ denote the $i$-th odd prime. For $k \geq 2$,

$$(p_k - 2) | \prod_{i=1}^{k-1} p_i(p_1 - 2).$$

6.5. *Remark.* As in the case of Pomerance's conjecture [16], our Conjecture 6.4 fails if there is a prime $p$ such that the smallest prime which is $\equiv 2 (\mathrm{mod}\ p)$ is also $\equiv 2 (\mathrm{mod}\ p^2)$. If Schinzel's Hypothesis $H_2$ holds, then there is no such $p$. For the convenience of the reader, Hypothesis $H_2$ essentially says that for each $n > 1$, there is at least one prime in the set $\{k, k+n, \ldots, k+n^2-n\}$ for each $k$ with $1 \leq k \leq n$, $(k, n) = 1$.

For the case of $\Phi_k$ for $k > 2$, one may construct a sequence like the $\{q_n\}$ in 6.2 by defining $q_{k,1} = k+1$ if $k+1$ is prime, $q_{k,2} = 2k+1$ if $2k+1$ is prime and then define for $n \geq 2$, $q_{k,n+1}$ = smallest prime $> q_{k,n}$ such that $(q_{k,n+1} - k)|(q_{k,1} \ldots q_{k,n})$ and safely assert that if the sequence $\{q_{k,n}, n = 1, 2, \ldots\}$, is infinite, then $N_k(m) \neq 1$ for any $m$. But for almost all $k$, the sequence $\{q_{k,n}\}$, $(n = 1, 2, \ldots)$ appears to be finite.

Indeed, in a letter to me dated May 5, 1988, P. Erdös wrote to say that ($k$ being arbitrary) if we start with $k+1$ and $2k+1$ as the first two terms of a sequence and define the third term $q$ to be the smallest prime $> 2k+1$ for which $(q-k)|(k+1)(2k+1)$, then he thinks that for almost all primes there is no such third term $q$ at all! Note that here $k$ is an arbitrary natural number and $k+1$, $2k+1$ need not be prime.

Hence, if $\ell_k = |\{q_{k,n}\}|$ = the length of the sequence $\{q_{k,n}\}$, then it is likely that for almost all $k$ we have $\ell_k = 2$. For example, $\ell_{18} = \ell_{30} = 2$, $\ell_{156} = 3$, $\ell_{178} = 4$. If however, $\ell_k = \infty$ for a given $k$, then arguing as before for $\Phi_2$, we can conclude that $N_k(m) \neq 1$ for any $m$ for that value of $k$.

We do not go here into the tricky question: if for a $k$, say $k = k_1$, $\ell(k_1)$ is finite, then does it necessarily mean that $N_{k_1}(m) = 1$ for some $m$?

6.6. *Remarks.* We do not know whether there are infinitely many integers $k > 0$ for which both $k+1$ and $2k+1$ are primes. The answer is "yes" if and only if the following holds:

6.7. *Conjecture.* There are infinitely many primes $p$ for which $p+1$ is twice a prime.

Compare this with Erdös Conjecture 4.9. Conjecture 6.7 is a companion to the

6.8. *Conjecture.* There are infinitely many primes $p$ for which $p-1$ is twice a prime. Assuming the truth of this, we can immediately prove.

6.9. *Conjecture (Katai [10]).* As $p$ runs over all the primes, $\lambda(p+1)$ takes both the values 1 and $-1$, each infinitely many times. Here $\lambda(n)$ is the Liouville function $=(-1)^{\Omega(n)}$, where $\Omega(n)$ denotes the total number of prime factors of $n$, multiplicity being taken into account.

I propose a companion to this:

6.10. *Conjecture.* As $p$ varies overall primes, $\lambda(p-1)$ takes each of the values $-1$ and $+1$ infinitely often.

Actually this is implied by Conjecture 6.5 which obviously says that there exist infinitely many primes $p$ and $q$ satisfying $p-1=2(q-1)$. This yields $\lambda(p-1)=\lambda(2)\lambda(q-1)=-\lambda(q-1)$.

## 7. Concluding remarks: some partition-theoretic problems

Many of the problems in the preceeding sections appeared within the last twentyfive years. They are mentioned here because I felt that they are very interesting and challenging and not too well known and certainly deserve the reader's attention.

As a completely new one, I mention the following (see my abstract #02T-11-17, vol. 23, no. 4, AMS Abstracts, 2002, issue 130, p. 639).

**7.1. Problem.** Let $n=\Pi p_i^{a_i}, i=1,\ldots,r$. Let $S=S(n)$ denote the set of all positive integers unitarily prime to $n$, that is, those natural numbers not divisible by any one of the $p_i^{a_i}, i=1,\ldots,r$. Let $p(S;n)$ denote the number of partitions of $n$ into distinct parts that belong to $S$. For example, 18 has four such partitions: 17+1, 15+3, 13+5, 11+7. Call an integer $n$ a champion with respect to $p(S;n)$ whenever $p(S,n)>p(S,m)$ for all $m<n$. What are the champions of $\{p(S;n)\}$?

(7.1.1) My conjecture is that they consist of all the primes and their powers with the sole exception of the integer 4. I further add that a similar conjecture holds for the case of partitions when the parts selected from the set S are allowed to repeat.

We may note that $\log p(S;n) \sim \pi 3^{-1/2}\left\{\prod_{i=1}^{r}(p_i^{a_i}-1)\right\}^{1/2}, (n\to\infty)$.

This follows from the following general theorem of Erdös [8]. Let $a_1<a_2<\ldots$ be an infinite sequence of integers of density $\alpha$ such that every sufficiently large $n$ can be expressed as the sum of different $a$'s. Then

$$\log p'(n)\sim c(\alpha n)^{1/2},\quad c=\pi\left(\frac{2}{3}\right)^{1/2},$$

and

$$\log P'(n)\sim c\,(\alpha n(2))^{1/2}.$$

Here $p'(n)$ and $P'(n)$ are the number of partitions of $n$ into $a$'s or distinct $a$'s respectively. We give a detailed account of $p(S; n)$ elsewhere.

Note also that the density of the set of all numbers unitarily prime to $n = \Pi p_i^{a_i}$ is $\varphi^*(n)/n$, where $\varphi^*(n)$ is the unitary totient given by $\varphi^*(n) = \Pi(p_i^{a_i} - 1)$.

7.2. *Remark.* The corresponding problem when $S(n)$ consists of numbers relatively prime to $n$ was mentioned by Erdös and Richmond [7]. They believe that the champions in this case are the primes and state, without proof, that this implies that between an integer $n$ and $n - n^{\frac{1}{2}+\varepsilon}$ there is a prime.

There is some connection between our conjecture (7.1.1) and the following conjecture, but we shall not go into details.

(7.2.1) *Conjecture.* For all sufficiently large $n$, between an integer $n$ and $n + n^{1/2}$, there is a prime power $p^h$ for some integer $h \geq 1$.

There seem to be no results in the literature on this conjecture and Professor K. Ramachandra confirmed this in a letter to the author.

The conjecture (7.2.1) is weaker than the corresponding conjecture for the case $h = 1$, but perhaps is no easier.

For the case $h = 1$ of (7.2.1), we recall that the best known result so far is due to R.C. Baker, G. Hartman and J. Pintz (The difference between consecutive primes - II, Proc. London Math. Soc., (3) 83 (2001), 532-562). They proved conjecture (7.3.1) in the case when $n^{1/2}$ is replaced by $n^{0.525}$ and $h = 1$.

We conclude this paper with a brief consideration of another partition function which we denote by $e^*(n)$. It has been introduced by Verma and myself [23] about two years ago and is defined by

$$\prod_{n=2}^{\infty}(1 - n^{-s}) = 1 + \sum_{n=2}^{\infty} e^*(n)n^{-s}, \qquad \mathrm{Re}\, s > 1,$$

Call a representation of $n > 1$ in the form $n = a_1 \ldots a_r$, where each $a_i > 1$, a product partition of $n$. The $a_i$'s are called its parts. Then, clearly, $e^*(n)$ represents the excess of the number of product partitions of $n$ into an even number of distinct parts over those into an odd number of such parts. The total number of such partitions of $n$ is denoted by $q^*(n)$. N.G. de Bruijn [4] studied this in detail.

If $n = p_1^{b_1} \ldots p_r^{b_r}$ is the prime factorization of $n$, we see that $e^*(n)$ and $q^*(n)$ depends on the exponents $b_1, \ldots, b_r$ only and not on the primes $p_1, \ldots, p_r$. Hence when $n$ is square-free and is the product of $k$ primes, we see that $e^*(n)$ is a function of $k$; call it $f(k)$. Similarly, $g(k)$ denotes the value of $q^*(n)$ when $n$ is product of $k$ distinct primes. In [4] de Bruijn obtained the result that if $L = \log k$, $L_2 = \log\log k$, $\log g(k) = k\big(L - L_2 - 1 + \frac{L_2+1}{L} + \frac{L_2^2}{2L^2} + O(\frac{L_2^3}{L^3})\big)$. From this we get

$$\limsup_{n\to\infty} \frac{\log|g(k)|}{k\log k} \leq 1. \tag{7.3}$$

Using the Euler pentagonal number theorem we easily see that $e^*(n)$ takes each of the values $-1, 0, 1$ infinitely many times. In fact if $q$ is any prime, then $e^*(q^m) = (-1)^k$

whenever $m = k(3k \pm 1)|2$, $k = 1, 2, \ldots$, and $= 0$ for other values of $m$. But $e^*(n)$ takes each of the values $-1, 0, 1$ for certain other values of $n$ also which are not of the form $q^m$; for example $e^*(6) = 0$. Actually, $f(k) = -1, 0$ or $-1$ for all $k \leq 4$.

Several problems have been raised in my paper with Verma [23]. Some of them are settled by Yifan Yang [28], using analytical methods, in a paper that appeared in 2001. However, many others remain. Before mentioning some of them, let me give a few numerical values of $f(k)$ below.

For $k = 1, 2, \ldots, 8$, $f(k)$ has the values $-1, 0, 1, 1, -2, -9, -9, 50$, respectively. $f(11)$, $f(12)$, $f(13)$ have respectively the values $-2180$, $-1773$, $-50533$. The values of $f(14)$, $f(15)$, $f(16)$, $f(17)$ are all positive, but $f(18)$, $f(19)$, $f(20)$ are all negative.

This illustrates the result that $f(k)$ is not monotonic with $k$ and changes sign – in fact infinitely often – as conjectured in our paper [23]. We proved that

$$\limsup_{k \to \infty} \left( |\log f(k)| \Big/ k \log k \right) = 1, \tag{7.4}$$

so that $\sup |f(k)|$ grows exponentially with $k$ and $|f(k)|$ is not uniformly bounded. Note that result (7.3) is weaker than (7.4).

Some problems and conjectures are the following:

(i) What is the density of the set of $n$ for which $e^*(n)$ takes each of the values $-1, 0, 1$?

7.5. *Remark.* Yang [28] showed that the density of the zeros of $f(k)$ is zero. Using an exponential sum estimate, he proved that the number of zeros of $f(k)$ that are $\leq x = O(x^{2/3})$. But the question remains whether $f(k) = 0$ infinite often. This question was also raised by H.S. Wolf ([28], p. 2).

(ii) Does $f(k)$ take any given value only a finite number of times? In particular is it true that except for 1 and $-9$, $f(k)$ takes no value more than once.

(iii) Is $k = 2$ the only case of $f(k)$ taking the value zero?

7.6. *Conjecture.*

(iv) The density $\delta$ of the set of $n$ for which $e^*(n)$ takes values $-1, 0$ or $1$ satisfies $0 < \delta < 1 - 6/\pi^2$, and for almost all square-free numbers $m$, $|f(m)| > 1$.

(v) Let $p^*(n)$ denote the number of product partitions of $n > 1$, with parts allowed to repeat. Then in 1999 [23], I made the

7.7. *Conjecture.* In every geometric progression ($a$, $r$ being integers) $a, ar, ar^2$, ($a \geq 1$, $r > 1$), there are infinitely many terms at which $p^*$ takes even values and infinitely many terms where $p^*$ takes odd values.

7.8. *Remark.* This is an extension of my 1966 conjecture [22], namely, if $p(n)$ denotes the number of unrestricted additive partitions of $n$, then

7.9. *Conjecture.* $p(n)$ takes each of even and odd values infinitely often in any given arithmetic progression.

Note that if $q$ is a prime, $p^*(q^n) = p(n)$.

Ken Ono [14] proved that Conjecture 7.9 holds for the even case and for some special forms in the odd case. Hence 7.7 holds for the even case when in the geometric progression $ar^n$, $a$ and $r$ are powers of the same prime.

Also, in view of the very recent work of M. Boylan and K. Ono [1], who showed that my conjecture for $p(n)$ holds for both the even and odd cases whenever the modulus $d$ of the arithmetic progression $\{a+md,\ m = 1, 2, \ldots\}$ is a power of 2, we can now state a little more regarding the conjecture for $p^*(n)$, namely that it holds for the geometric progression $\{ar^n,\ n = 1, 2, \ldots\}$ for both the even and odd cases whenever $a$ and $r$ are powers of the same prime $q$, and $r$ has the form $q^{2^k}$, $k = 1, 2, \ldots$.

At the November 1999, Conference on Symbolic Computation, Number Theory, Special Functions, Physics and Combinatories held in Florida, the author announced this conjecture on $p^*(n)$ and in response to a query from the audience and in the tradition of Paul Erdös, offers $500 (Canadian) for the first correct and complete resolution of the same.

## References


[1] M. Bolyan and Ken Ono, Parity of the partition function in arithmetic progressions II *Bull. London Math. Soc.* **33** (2001) 558–564

[2] R. D. Carmichael, Note on Euler's $\phi$-function *Bull. Amer. Math. Soc.* **28** (1922) 109–110

[3] S. Chowla, *J. Indian Math. Soc.* **18** (1930) 230

[4] N. G. de Bruijn, *Asymptotic Methods in Analysis* Corrected third edition, Dover, New York 1981

[5] P. Erdös, Some remarks on Euler's $\varphi$-function *Acta Arith.* **4** (1958) 10–19

[6] P. Erdös and R. L. Graham, Old and new problems and results in combinatorial number theory Monographie No. 28, *L;Einseignement Mathematique*, Université de Geneva (1980)

[7] P. Erdös and B. Richmond, On partitions of $N$ into summands coprime to $N$ *Aequationes Mathematicae* **18** (1978) 178–196

[8] P. Erdös, On an elementary proof of some asymptotic formulae in the theory of partitions *Annals of Math.* **43** (1942) 437–450

[9] G. E Hardy and M. V. Subbarao, On a modified Pillai problem and related questions *Amer. Math. Monthly* **109** (2002) 554–559

[10] I. Katai, Research problems in number theory *Pub. Math. Debrecen* **24** (1977) 263–276

[11] V. L. Klee, Is there an $n$ for which $\phi(x) = n$ has a unique solution? *Amer. Math. Monthly* **76** (1969) 288–289

[12] P. Masai and A. Valette, A lower bound for a counterexample to Carmichael's conjecture *Bollettino delta Unione Matematica Italiana, Serie VI*, **1-A** (1982) 313–316

[13] L. J. Mordell, *Diophantine equations* London Academic Press (1969)

[14] K. Ono, Parity of the partition function in arithmetic progressions *J. reine angew math.* **472** (1966) 1–15

[15] S. S. Pillai, Question 1490 *J. Indian Math. Soc.* **18** (1930) 230

[16] C. Pomerance, On Carmichael's conjecture *Proc. Amer. Math. Soc.* **43** (1974) 297–298

[17] A. Schinzel, Remarks on the paper "Sur certaines hypothèses concernant les nombres premiers" *Acta Arith.* **7** (1961) 1–8

[18] A. Schinzel and W. Sierpinski, Sur certaines hypothèses concernant les nombres premiers *Acta Arith.* **4** (1958) 185–208

[19] B. M. Stewart, *Theory of Numbers* Second Edition, Mac Millan, New York (1964)

[20] B. M. Stewart and W. A. Webb, Sums of fractions with bounded numerators *Canadian J. Math.* **18** (1966) 999–1003

[21] E. G. Straus and M. V. Subbarao, On the representation of fractions as sum and difference of three simple fractions *Proc. Seventh Manitoba Conference on Numerical Mathematics and Computing*, Congress Numerantium XX (1977)

[22] M. V. Subbarao, Some remarks on the partition function *Amer. Math. Monthly* **73** (1966) 851–854

[23] M. V. Subbarao and A. Verma, Some remarks on a product expansion – an unexplored partition function *Number Theory, Special Functions, Physics and Combinatorics* (F G Garvan and M E H Ismail), Kluwer Academic Publishers (2001) 267–283

[24] M. V. Subbarao and L. W. Yip, Carmichael's conjecture and some analogues *Proc. Int. Number Theory Conf. Univ.* Laval 1987 (De. Koninck and Levesque) Walter de Gruyter, Berlin (1989) 928–941

[25] M. V. Subbarao and L. W. Yip, On Sierpinski's conjecture concerning the Euler totient *Canadian Math. Bull.* **34** (1991) 401–404

[26] R. C. Vaughan, On a problem of Erdös, Straus and Schinzel *Mathematika* **17** (1970) 193–198

[27] William Webb, Expressing rationals as a sum of a small number of unit fractions *Casopis Pest. Mat.* **105** (1980) 345–349

[28] Yifan Yang, On a multiplicative partition function *The electronic J. Combinatorics* **8** (2001) 1–14

[29] L. W. Yip, *On Carmichael type problems for the Schemmel totients and some related questions* Doctoral Thesis, Univ. of Alberta (1989)

Proc. Int. Conf. – Number Theory No. 1, 2004, pp. 135–141.

# Some Generalisations of Brauer–Rademacher Identity and Menon Sum Identity

V. V. Subrahmanya Sastri

203, SMR Square, 6-2-97/1, New Boiguda, Secunderabad 500 003, India

*Dedicated to octogenarian Professor M. V. Subbarao*

## 1. Introduction

The Brauer–Rademacher Identity [2] was given for the Euler totient function. Cohen [3–6] gave new proofs and some generalisations of these. Mc Carthy [8] also gave a proof and a generalisation.

In this humble paper, we prove a generalisation of a theorem of Subbarao [9] and use it to obtain some generalisations of the Brauer–Rademacher Identity. We also obtain some generalisations of the Kesava Menon sum identity [7]. These results are related to the Klee's totient function defined by

$$\phi_k(n) = \text{Card}\{a \bmod n : (a, n)_k = 1\},$$

where for any three positive integers $a, k, n$ the notation $(a, k)_n$ denotes the greatest $k$th power common divisor of $a$ and $n$, and also to the generalised Nagell totient defined by

$$\theta_k(m_1, m_2; n) = \text{Card}\{a \bmod n : (a, n)_k = (m_1 - a, n)_k = (m_2 - a, n)_k = 1\},$$

where $m_1, m_2, k, n$ are any four positive integers.

## 2. Preliminaries

Recall that a complex valued arithmetic function $f(r)$ is said to be multiplicative if $f(1) = 1$ and $f(r_1 r_2) = f(r_1) f(r_2)$ whenever $(r_1, r_2) = 1$. A complex valued arithmetic function in two variables $s, r$ is said to be *partially multiplicative in r* if $f(1, 1) = 1$ and for any $s$, $f(s, r_1 r_2) = f(s, r_1) f(s, r_2)$ whenever $(r_1, r_2) = 1$. Finally, we say that a complex valued arithmetic function in two variables $s, r$ is *multiplicative in both s and r* if $f(1, 1) = 1$ if $f(1, 1) = 1$ and $f(s_1 s_2, r_1 r_2) = f(s_1, r_1) f(s_2, r_2)$ whenever $(r_1 s_1, r_2 s_2) = 1$.

The following lemma is easily proved.

**Lemma 1.** *When $f(r)$ and $g(r)$ are multiplicative functions let $h(n, r)$ be the arithmetic function defined by*

$$h(n, r) = \sum_{\substack{d^k \mid r, \\ (d^k, n)_k = 1.}} f(d^k) g\left(\frac{r}{d^k}\right). \tag{1}$$

*Then $h(n, r)$ is a multiplicative function in both $n$ and $r$ and, further, is partially multiplicative in $r$.*

When $k = 1$ the above lemma reduces to the lemma of Subbarao [9].

## 3. The main result

The main result of this note is the following theorem, which reduces to the theorem of Subbarao [9] when $k = 1$. Here we have written $\mu(n)$ to denote the Möbius function.

**Theorem 1.** *Let $k$ be a positive integer and suppose that $f(r)$ is a multilplicative function such that $f(p^{lk}) = f(p^{(l+1)k})$ for all $l \geq 1$ and for each prime $p$. Let $h(r)$ be a multiplicative function such that $h(p^k) = f(p^k) - 1$ for every prime $p$. When $r$ and $n$ are $\geq 1$ we then have*

$$\sum_{\substack{d^k \mid r, \\ (d^k, n)_k = 1.}} f(d^k) \mu_k\left(\frac{r}{d^k}\right) = \mu_k(r) \mu_k\left(\frac{r}{t}\right) h_k\left(\frac{r}{t}\right), \tag{1}$$

*where $t = (r, n)_k$ and $\mu_k(n) = \mu(n^{1/k})$ when $n$ is a $k$th power and is 0 otherwise.*

*Proof.* It follows from Lemma 1 of Section 2 that both sides of (1) are multiplicative functions of both $n$ and $r$. On writing $F(n, r)$ and $H(n, r)$ to denote the left and right hand sides of (1), it thus suffices to verify that

$$F(p^b, p^a) = H(p^b, p^a), \tag{2}$$

when $p$ is prime and $a$, $b$ non-negative integers. Let $l$ and $m$ be non-negative integers such that $lk \leq a < (l+1)k$, $mk \leq b < (m+1)k$. On evaluating $F(p^b, p^a)$ we easily see that

$$F(p^b, p^a) = \sum_{0 \leq s \leq l = [a/k],\ \min(sk, b) < k} f(p^{sk}) \mu_k(p^{a-sk}). \tag{3}$$

The right hand side of (3) is 1 for any $b$ when $a = 0$, is $f(p^k) - 1$ for any $b$ when $a = k$, and is 0 for all other values of $(a, b)$. On using the hypothesis on $f$ and $h$ to evaluate $H(p^b, p^a)$ we then conclude (2).

## 4. Preliminary applications

As immediate corollaries to Theorem 1 of Section 3 we have the following results.

We set $f(r) = 1$ for all $r$ and set $h(r) = [1/r] = 1$ or 0 according as $r = 1$ or $r > 1$. We then obtain that

$$\sum_{\substack{d^k \mid r, \\ (d^k, n)_k = 1.}} \mu_k\left(\frac{r}{d^k}\right) = \left[\frac{1}{r}\right]. \tag{1}$$

On setting $f(r) = 2^{\omega(r)}$, where $\omega(r)$ is number of distinct prime factors of $r$, and setting $h(r) = |\mu_k(r)|$ we obtain

$$\sum_{\substack{d^k \mid r, \\ (d^k, n)_k = 1.}} 2^{\omega(d)} = \mu_k(r)\mu_k(r/t)|\mu_k(r/t)|. \tag{2}$$

Let $\phi_{k,\lambda}(r)$ be the number of ordered sets $\{a_1, a_2, \ldots, a_\lambda\}$ each $a_i$ chosen from a complete residue system ( mod $r$) such that $(r, a_1, a_2, \ldots, a_\lambda)_k = 1$ (this is the Jordan's totient generalisation of Klee's totient). Then, on choosing $f(r) = r^\lambda/\phi_{k,\lambda}(r)$ and $h(r) = 1/\phi_{k,\lambda}(r)$, we have

$$\sum_{\substack{d^k \mid r, \\ (d^k, n)_k = 1.}} \frac{d^{k\lambda}\mu_k(r/d^k)}{\phi_{k,\lambda}(d^k)} = \frac{\mu_k(r)\mu_k(r/t)}{\phi_{k,\lambda}(r/t)}, \tag{3}$$

where $t = (r, n)_k$. This is a generalisation of Cohen's result [6]. When $k = 1$ these give the results of Subbarao [9].

## 5. A generalisation of Nagell's quotient and an allied identity

Let $m_1, m_2, n$ and $k$ be positive integers. We say that a prime number $p$ is of type 1 if $p|(n, m_1, m_2)$, of type 2 if $p|n$ and $p$ divides just one of $m_1, m_2, m_1 - m_2$ and of type 3 if $p|n$ and $p$ does not divide any of $m_1, m_2, m_1 - m_2$ (see Subbarao and Subrahmanya Sastri [10]). Further, we write $\omega_1(n)$, $\omega_2(n)$, $\omega_3(n)$ to denote the number of distinct prime divisors of the types 1, 2 and 3 respectively which divide $n$. Finally, we let

$$\mu_k(m_1, m_2; n) = (-1)^{\omega_1(n^{1/k})}(-2)^{\omega_2(n^{1/k})}(-3)^{\omega_3(n^{1/k})},$$

when $n$ is a $k$ th power and 0 otherwise.

We have the formula

$$\theta_k(m_1, m_2; n) = \prod_{p_1}\left(1 - \frac{1}{p_1^k}\right)\prod_{p_2}\left(1 - \frac{2}{p_2^k}\right)\prod_{p_3}\left(1 - \frac{3}{p_3^k}\right)$$

where $p_1$, $p_2$ and $p_3$ run over those prime divisors of types 1, 2 and 3 respectively, whose $k$th powers divide $n$ and $\theta_k(m_1, m_2; n)$ is as defined in Section 1. The case $k = 1$ of the above result is discussed in Subbarao and Subrahmanya Sastri [10] with many applications. We also note that

$$\theta_k(m_1, m_2; n) = n,$$

when $n$ is $k$-free and that

$$\theta_k(m_1, m_2; n_1 n_2) = n_1 \theta_k(m_1, m_2; n_2),$$

whenever $n_1$ is $k$-free and $n_2$ is $k$-full.

We now present an analogue of Theorem 1 of Section 3.

**Theorem 1.** *Let $k$ be a positive integer and suppose that $f(r)$ is a multilplicative function such that $f(p^{lk}) = f(p^{(l+1)k})$ for all $l \geq 1$ and for each prime $p$. Let $h(r)$ be a multiplicative function such that $h(p^k) = [f(p^k) - 1]/u$ for any prime $p$ of the type $u$, $u = 1, 2, 3$. When $r$ and $n$ are $\geq 1$ we then have*

$$\sum_{\substack{d^k | r, \\ (d^k, n)_k = 1}} f(d^k) \mu_k \left(m_1, m_2, \frac{r}{d^k}\right) = \mu_t(r) \mu_k \left(m_1, m_2; \frac{r}{t}\right) h\left(\frac{r}{t}\right), \tag{1}$$

*where $t = (r, n)_k$.*

The proof runs along the same lines as that of Theorem 1 of Section 3 and is not given here for this reason.

**Corollary.** On setting $f(r) = r/\theta_k(m_1, m_2; r)$ and $h(r) = 1/\theta_k(m_1, m_2; r)$ we obtain

$$\sum_{\substack{d^k | r, \\ (d^k, n)_k = 1}} d^k \mu_k \left(m_1, m_2, \frac{r}{d^k}\right) = \frac{\mu_k(r) \mu_k \left(m_1, m_2; \frac{r}{t}\right)}{\theta_k(m_1, m_2; \frac{r}{t})}, \tag{2}$$

where $t = (r, n)_k$.

The identities (1) and (2) are generalisations of the Brauer–Rademacher identity.

## 6. Menon sum identity analogues

Kesava Menon [7] has obtained that

$$\sum_{a \bmod n, (a,n)=1} (a - 1, n) = \phi(n) \tau(n),$$

where $\tau(n)$ is the number of divisors of $n$. Our aim in this section is to obtain some generalisations of the above result. To this end we recall or state the following results which hold.

We have the following identities.

$$\sum_{d | n} \phi(d) = n, \tag{1}$$

$$\sum_{d^k | n} \phi_k(n | d^k) = \sum_{d | n} \phi_k(d) \lambda_k(n/d) = n, \tag{2}$$

where $\lambda_k(n) = 1$ if $n$ is a $k$th power and is 0 otherwise, and finally we have

$$\sum_{d^k | n} \phi_k(d^k) = n, \quad \text{when } n \text{ is a } k\text{th power.} \tag{3}$$

We next have the following lemmas analogous to Theorem 5.3.2 of Apostol [1].

**Lemma 1.** *When $d$ is a divisor of $n$ and $(t,d)_k = 1$ then there are $\phi_k(n)/\phi_k(d)$ integers that are congruent to $t(\bmod\ d)$ in $R^*(k,n)$.*

**Lemma 2.** *When $d > 1$ is a divisor of $n$ and $(t,d)_k = (m_1 - t, d) = (m_2 - t, d) = 1$ then there are $\theta_k(m_1, m_2, ; n)/\theta_k(m_1, m_2, ; d)$ integers that are congruent to $t(\bmod\ d)$ in $R^*(k, m_1, m_2; n)$ provided that $(t,n)_k$, $(m_1 - t, n)$ and $(m_2, -t; n)_k$ do not divide $d$ (for $d > 1$).*

The proof of Lemma 2 relies on the following lemma.

**Lemma 3.** *When $a$ is in $R^*(m_1, m_2; n)$ and $t$ is in $R^*(m_1, m_2; d)$ where $d > 1$ is a divisor of $n$ and $a$ is congruent to $t(\bmod\ d)$, then $(t,n)_k$, $(m_1 - t, n)$ and $(m_2, -t; n)_k$ do not divide $d$.*

*Proof.* To see this it is enough to note that if $a$ is congruent to $t(\bmod\ d)$ under the hypothesis here then $(t,n)_k$ does not divide $d$ and similarly $(m_1, -t; n)_k$ and $(m_2, -t; n)_k$ do not divide $d$.

In fact when $a$ is congruent to $t(\bmod\ d)$, $t$ in $R^*(m_1, m_2; d)$and $a$ in $R^*(m_1, m_2; n)$, we must have $(t,n)_k = \delta^k$ which implies that $\delta^k$ divides $t$. Since $(a,n)_k = 1$ let $a = t + sd$, say, $(t + sd, n)_k = 1$, $\delta | t$ implies $\delta^k$ is not a divisor of $d$.

Similarly $(m_1 - a, n)_k = 1$ and $(m_2 - a, n)_k = 1$ respectively imply that $(m_1 - t; n)_k$ and $(m_2 - t; n)_k$ do not divide $d$.

Lemma 2 now follows along the same lines as Theorem 5.3.2. of Apostol [1]. Lemma 2 allows us to deduce the following analogues of the Menon sum identity.

**Theorem 1.** *We have*

$$\sum_{a \in R^*(k,n)} (a - 1, n) = \phi_k(n) t_k(n),$$

*where $t_k(n) = \sum_{d|n} \phi(d)/\phi_k(d^k)$.*

**Theorem 2.** *We have*

$$\sum_{a \in R^*(k,n)} (a - 1, n)_k = \phi_k(n) \tau_k(n),$$

*where $\tau_k(n)$ is the number of $k$th power divisors of $n$.*

**Theorem 3.** *We have*

$$\sum_{a \in R^*(k, m_1, m_2; n)} (a - 1, n) = \theta_k(m_1, m_2; n) T_k(m_1, m_2; n),$$

*where*

$$T_k(m_1, m_2; n) = \sum_{d \in S_k(m_1, m_2, n)} \frac{\phi(d)}{\theta_k(m_1, m_2; d)}$$

*and $S_k(m_1, m_2, n) = \{d : d|n$ and $(m_1 - 1, n)$ and $(m_2 - 1, n)$ do not divide $d$, $d > 1\}$.*

**Theorem 4.** *We have*

$$\sum_{a\in R^*(k,m_1,m_2;n)} (a-1,n)_k = \theta_k(m_1,m_2;n)\mathcal{I}_k(m_1,m_2;n),$$

*where* $\mathcal{I}_k(m_1,m_2;n)$ *denotes*

$$\sum_{d\in S_k(m_1,m_2,n)} \frac{\phi_k(d^k)}{\theta_k(m_1,m_2;d^k)}.$$

We shall prove only Theorems 3 and 4 stated above.

*Proof of Theorem 3.* We have on using the relation (1) that

$$\sum_{a\in R^*(k,m_1,m_2;n)} (a-1,n) = \sum_{a\in R^*(k,m_1,m_2;n)} \sum_{d|(a-1,n)} \phi(d).$$

On interchanging the summations, we see that the left side of the above relation is the same as

$$\sum_{d|n} \phi(d) \sum_{a\in R^*(k,m_1,m_2;n), a=1 (\text{mod } d)} 1.$$

On using Lemma 2 for $t = 1$ we thus see that the above expression is the same as

$$\sum_{d\in S_k(m_1,m_2;n)} \phi(d)\theta_k(m_1,m_2;n)/\theta_k(m_1,m_2;d) = \theta_k(m_1,m_2;n)T_k(m_1,m_2;n).$$

This completes the proof of Theorem 3.

*Proof of Theorem 4.* Let $n = N^k m$, where $m$ is $k$th power free. Then $d^k$ divides $n$ if and only if $d$ divides $N$. Thus we have on using the relation (2) that

$$\sum_{a\in R^*(k,m_1,m_2;n)} (a-1,n)_k = \sum_{a\in R^*(k,m_1,m_2;n)} \sum_{d^k|(a-1,N^k)_k} \phi_k(d^k).$$

On interchanging the summations, we see that the left side of the above relation is the same as

$$\sum_{d^k|n} \phi_k(d^k) \sum_{a\in R^*(k,m_1,m_2;n), a=1 (\text{mod } d)} 1.$$

On using Lemma 3 with $d^k$ in place of $d$ and $t = 1$ we thus see that the above expression is the same as

$$\sum_{d^k\in S_k(m_1,m_2;n)} \phi_k(d^k)\theta_k(m_1,m_2;n)/\theta_k(m_1,m_2;d^k) = \theta_k(m_1,m_2;n)\mathcal{I}_k(m_1,m_2;n).$$

This completes the proof of Theorem 4.

The author thanks Dr. D. Surya Ramana for carefully going through an earlier version of the manuscript and for his comments.

## References

[1] T. M. Apostol, *An Introduction to Analytic Number Theory*, Springer Student Edition, Narosa Publishing Company (1990)

[2] A. Brauer and H. Rademacher, Aufgabe 31, jber Deutsch, *Math Verein*, **35** (1926) 94–95

[3] E. Cohen, The Brauer–Rademacher Indentity, *American Math. Monthly*, **67** (1960) 30–33

[4] E. Cohen, Arithmetical Inversion Formulas, *Canadian Journal of math.* **12** (1960) 399–409

[5] E. Cohen, Representation of Even Functions $(\bmod r)$ Cauchy Products, *Duke Math. Journal.* **26** (1959) 165–182

[6] E. Cohen, A Trignometric Sum, *The Math. Student.* **28** (1960) 29–32

[7] P. Kesava Menon, On the sum $\sum_{(a,n)=1}(a-1,n)$, *J. Indian. Math. Soc.*, (2nd Series) **26** (1965)

[8] P. J. Mc Carthy, Some Remarks on Arithmetical Identities, *American Math. Monthly*, **72** (1965) 135–138

[9] M. V. Subbarao, The Brauer–Rademacher Identity, *American Math. Monthly*, **67** (1960) 539–548

[10] M. V. Subbarao and V. V. Subrahmanya Sastri, On extensions of Nagell's totient function and some applications, *Ars Combinatoria*, **62** (2002) 79–96

Proc. Int. Conf. – Number Theory No. 1, 2004, pp. 143–153.

# Central Extensions of a $p$-adic Division Algebra

B. Sury

Statistics and Mathematics Unit, Indian Statistical Institute, 8th Mile Mysore Road, Bangalore 560 059, India
e-mail: sury@isibang.ac.in

**Subject Classification:** 20 G 25

## 1. Introduction

Let $k$ be a finite extension of the field $\mathbf{Q}_p$ of $p$-adic numbers and suppose $D$ is a finite-dimensional central division algebra over $k$.

Then, the group $G = SL_1(D)$ consisting of elements of reduced norm 1 in $D$ acquires a topology from $k$ and is a compact, totally disconnected (i.e., a profinite) group. We are interested in finding the possible (topological) central extensions

$$1 \to A \to E \to G \to 1.$$

General nonsense tells us that the set of central extensions is determined by a group, denoted $H^2(G, \mathbb{R}/\mathbb{Z})$, which, in this case, is known to be finite by some deep work of Raghunathan on the congruence subgroup problem ([R]). It is expected (although unknown as yet) that, barring a handful of exceptions, $H^2(G, \mathbb{R}/\mathbb{Z}) \cong \mu(k)_p$, the finite cyclic group of $p$-th power roots of unity in $k$. In 1988, Gopal Prasad & M. S. Raghunathan proved ([PR]) that $H^2(G, \mathbb{R}/\mathbb{Z})$ is a finite cyclic group containing an isomorphic copy of $\mu(k)_p$ and is trivial if $\mu(k)_p$ is trivial.

These results were sufficient for their original motivation to solve the so-called metaplectic problem which comes up in the congruence subgroup problem. However, the general computation of $H^2$ is still open.

Our aim here is to stretch the method of [PR] and study the $p^2$-torsion in $H^2(G, \mathbb{R}/\mathbb{Z})$ with a view to proving that if $H^2$ has an element of order $p^2$, then $k$ contains a primitive $p^2$-th root of unity. The computations are rather cumbersome, and we carry them out fully only in a special case when $p = 3$ and $D$ is the quaternion division algebra although we have partial results in more generality.

One probably needs new ideas along with the work of [PR] if one wants to compute $H^2$ in general. Perhaps, on the other hand, the seminal work of Lazard ([L]) on compact $p$-adic Lie groups has not been exploited sufficiently enough.

## 2. Basic structure of $D$

The structure of $p$-adic division algebras had been investigated by C. Riehm in [Ri]. Let us briefly recall some details.

$D$ contains $R$, its maximal compact subring which, in turn, contains a unique (two-sided) maximal ideal $P$. The group $G = SL(1, D)$ of elements of reduced norm 1 in $D$, is a profinite group which is normal in $D^*$.

$G$ admits a filtration $G_i = \{g \in G : g \equiv 1 \bmod P^i\}$ for $i \geq 1$. In fact, $G_i$ are all normal in $D^*$. $G_1$ is a pro-$p$ group and $G/G_1$ is a finite, cyclic group of order prime to $p$.

For $i, j \geq 1$, we have $[G_i, G_j] \subseteq G_{i+j}$. In particular, $G_i/G_{i+1}$ is an abelian group for $i \geq 1$.

By local class field theory, there exists a uniformising parameter $\pi$ in $P$ which normalises the maximal unramified extension $K$ of degree $d$ over $k$. Also, the automorphism of $K$ given by the conjugation by $\pi$ generates the Galois group $\mathrm{Gal}(K/k) \cong \mathrm{Gal}\,(F/f)$ where $F$, $f$ are the residue fields of $K, k$ respectively.

Moreover, the non-zero elements $F^*$ can be identified with $\mu(K)_{\text{tame}}$, the cyclic group of prime-to-$p$ roots of unity in $K$.

Each element $g \in R$ can be uniquely expressed as $g = g_0 + \sum_{n \geq 1} g_n \pi^n$ with $g_n \in \mu(K)_{\text{tame}} \cup \{0\}$.

A little computation shows also that the abelian group $G_i/G_{i+1}$ can be identified via the map $\rho_i : 1 + \sum_{n \geq i} g_n \pi^n \mapsto g_i$ with $F(i)$ which is either $E := \{x \in F : Tr_{F/f}(x) = 0\}$ or the whole of $F$ according as $d|i$ or $d \nmid i$.

It is also quite easy to show that $G = G_1(G \cap \mu(K)_{\text{tame}})$.

Note that each $F(i)$ is a module for $G \cap \mu(K)_{\text{tame}}$ under the action $\phi \cdot x = \frac{\phi}{\sigma^i(\phi)} x$ where we have identified $\mu(K)_{\text{tame}}$ with $F \setminus \{0\}$. A consequence of Hilbert's theorem 90 is that there is a nontrivial homomorphism (of modules) from $F(i)$ to $F(j)$ if, and only if, $i \equiv j \bmod d$.

## 3. Conditions for roots of unity

It is easy to write down necessary and sufficient conditions for $k$ to contain a primitive $p^2$-th root of unity.

Recall that $k \subset K \subset D$ and $\pi$ is a uniformising parameter in $D$ normalising $K$. Let $e$ denote the ramification index of $k$ over $\mathbb{Q}_p$ and $d$ denote the degree of $D$ over $k$. Then $\pi^d$ is a uniformising parameter for $k$ and, one can expand $p$ over $k$ as

$$p = \theta \pi^{de} + \theta_d \pi^{de+d} + \cdots$$

for some $\theta$'s in $f$.

Now, well-known properties of the $p$-th power map (see [M], P. 167–168) tells us that $k$ has a primitive $p^2$-th root of unity if, and only if, $p(p-1)$ divides $e$ and there exists some $Y = 1 + Y_{de/p(p-1)} \pi^{de/p(p-1)} + Y_{de/p(p-1)+d} \pi^{de/p(p-1)+d} + \cdots$ such that $Y_{de/p(p-1)} \neq 0$ and such that $Y^{p^2} \equiv 1 \bmod \pi^{de/p(p-1)+de+d}$.

This gives (using the expression of $p$) certain polynomial equations in the $Y_i$'s with coefficients as some $\theta$'s. We get finitely many polynomials over $f$.

Our aim is, therefore, to deduce the simultaneous solvability of these equations by somehow getting information over $f$ that can be derived from the assumption that $H^2$ has $p^2$-torsion.

It should be pointed out that the corresponding calculation with $p$ in place of $p^2$ is much easier and was carried out in [PR].

## 4. Strategy of studying $H^2$ (after [PR])

In this section, we recall the basic method as well as the results of Prasad & Raghunathan from [PR] which we shall be using.

*Fact* 1. Using the fact that $G$ is profinite, it is easy to deduce that $H^2(G, \mathbb{R}/\mathbb{Z}) \cong H^2(G, J)$ where $J$ is the subgroup of $\mathbb{Q}/\mathbb{Z})$ consisting of $p$-power order (considered with the discrete topology).

We shall be using the Hochschild-Serre spectral sequence for the situation $G_{i-1}/G_i \leq G/G_i$ for $i > 1$.

A very useful property is that given a central extension $C \subseteq E \to A$ where $A$ is abelian is that one has a lifted 'commutator' map; if $a, b \in A$, then for arbitrary lifts $x, y$ of $a, b \in E$, the commutator $[x, y] = xyx^{-1}y^{-1}$ lands inside $C$ and, is independent of the lifts. One often writes $[a, b]$ for this element of $C$. When $C$ is a divisible group (like our $J$), the extension is 'trivial' if, and only if, $E$ itself is abelian.

In our case, we shall use it for $G_i/G_j$ which is abelian if $2i \geq j$.

*Fact* 2. $H^2(G, J) \overset{\lim}{=\!\!\rightarrow} H^2(G/G_i, J)$ and, the 'inflation' maps $\inf(i) : H^2(G/G_i, J) \to H^2(G, J)$ are injective if, and only if, $d \not| i$. Moreover, $H^2(G, J)$ is the union of the images $H^2(G)_i$ over all $i$ under the inflation maps; $H^2(G)_i$ is an increasing filtration.

If $d|i$, then there is a natural identification of Ker inf($i$) with the vector space $E$ of elements of trace zero in $F$ over $f$.

Thus, one needs to compare $H^2(G)_i$ and $H^2(G)_{i-1}$ for various $i$.

*Fact* 3. If $d \not| i$, $H^2(G)_i = H^2(G)_{i-1}$.

In fact, one can show that the same equality holds if $dp \not| i$; this uses some commutator identities due to P. Hall which are valid in any group.

(P. Hall) In any group $G$, for elements $a, b, c$ one has

$$[[a, b], {}^b c]][[b, c], {}^c a]][[c, a], {}^a b]] = 1.$$

Look at the inflation maps

$$H^2(G/G_{r-1}, J) \to H^2(G/G_r, J) \to H^2(G, J).$$

If $d|r$, since the inflation $\inf(r)$ is not injective, it is necessary to know when some $c \in H^2(G/G_r, J)$ inflates in $H^2(G, J)$ to an element which comes from $H^2(G/G_{r-1}, J)$. This happens if $c$ restricts to the trivial extension over the subgroup $G_{r-1}/G_r$. More precisely, using the Hochschild-Serre sequence corresponding to $G_{r-1}/G_r \lhd G/G_r$, we see that:

*Fact* 4. $c$ comes from $H^2(G/G_{r-1}, J)$ if, and only if, it is in $\mathrm{Ker}(H^2(G/G_r, J) \to H^2(G_{r-1}/G_r, J)) \cap E_\infty^{1,1}$.

This is understood better as follows.

For any $c$ in $H^2(G/G_r, J)$, let $J \subseteq E \to G/G_r$ denote the corresponding central extension. The $E_2^{1,1}$-term is

$$\begin{aligned} H^1(G/G_{r-1}, H^1(G_{r-1}/G_r, J)) &= H^1(G_1/[G_1, G_1], \mathrm{Hom}\,(F(r-1), J))^{G/G_1} \\ &= \mathrm{Hom}\,(F(1), \mathrm{Hom}\,(F(r-1), J))^{G \cap \mu(K)_{\mathrm{tame}}}. \end{aligned}$$

As we noted, this is nontrivial only if $d|r$. If $d|r$, this is just the set of all (equivariant) bimultiplicative maps from $F(1) \times F(r-1)$ to $f_0$ where we shall write $f_0$ for the prime field $\mathbb{Z}/p$.

In fact, it is easy to write down all its elements. These are the maps $F(1)\times F(r-1) \to f_0$; $(X, Y) \mapsto Tr_{F/f_0}(\lambda X\sigma(Y))$ for some $\lambda \in F$. With this identification, it follows that $c$ comes from the previous level if, and only if, the corresponding $\lambda$ has trace zero over $f$ i.e., we get:

*Fact* 5. $c$ in $H^2(G/G_r, J)$ is the inflation of an element of $H^2(G/G_{r-1}, J)$ if, and only if, there exists $\lambda \in F$ of trace zero over $f$ such that $c$ can be 'regarded' as (this is the image in the $E_2^{1,1}$-term) the map from $F(1) \times F(r-1)$ to $f_0 = \mathbb{Z}/p$ given by

$$\wedge^c(X, Y) = Tr_{F/f_0}(\lambda X\sigma(Y)) \ \forall\, X \in F(1), Y \in F(r-1).$$

It should be noted that the map above is induced by the 'commutator' map from $E_1 \times E_{r-1}$ to $J$ where $E_i$ is the inverse image of $G_i/G_r$ in $E$.

More generally, from P. Hall's identity, one can easily show that $E_i$ commutes with $E_{r-i+1}\ \forall\, i \leq r$. Thus the central extension splits over $G_i/G_r$ whenever $2i > r$. Thus:

*Fact* 6. If $r > 2$ and $\epsilon < \frac{r}{2}$ then the restriction map $H^2(G/G_r, J) \to H^2(G_{r-\epsilon}/G_r, J)$ is the zero map.

Once again, if $\epsilon < \frac{r}{2}$, we can consider the Hochschild-Serre sequence corresponding to $G_{r-\epsilon}/G_r \lhd G/G_r$. By the above, ($E_\infty^{0,2}$ - term is zero and so) we have a homomorphism

$$H^2(G/G_r, J) \to E_\infty^{1,1} \subseteq E_2^{1,1} = H^1(G/G_{r-\epsilon}, \mathrm{Hom}\,(G_{r-\epsilon}/G_r, J)).$$

One can show that the above-mentioned image in the $E_2^{1,1}$-term actually comes from $(H^1(G/G_{\epsilon+1}, \mathrm{Hom}\,(G_{r-\epsilon}/G_r, J))$. More precisely, :

*Fact* 7. With $\epsilon < r/2$, the image in $E_2^{1,1}$ is contained in the image of

$$\inf l : H^1(G/G_{\epsilon+1}, \mathrm{Hom}\,(G_{r-\epsilon}/G_r, J)) \to E_2^{1,1}.$$

A key point (discovered in [PR]) is that one can describe these 1-cocycles very explicitly. We describe this now.

*Fact* 8. Let $\epsilon, s, t$ be positive integers s.t. $\epsilon \leq \min.(de, \frac{1}{2}dt)$ and $s \geq \epsilon + 1$. Let $f = \left[\frac{\epsilon-1}{d}\right]$. For $(\lambda_0, \ldots, \lambda_f) \in F^{f+1}$,

$$Z_{(\lambda_0,\ldots,\lambda_f)}(a)(b) = Tr_{F/f_0}\left\{\sum_{0\leq u\leq f}\ \sum_{\ell+m\leq\epsilon-du} \lambda_u(\ell)a_\ell\sigma^\ell(a'_m)\sigma^{\ell+m}(b_{dt-du-\ell-m})\right\}$$

(here $a = \sum a_\ell\pi^\ell \in D_1/D_s, b = \sum b_m\pi^m \in D_{dt-\epsilon}/D_{dt}$) is a $F$-invariant 1-cocycle on $D_1/D_s$ with values in Hom $(D_{dt-\epsilon}/D_{dt})$; these restricted to $G_1/G_s \times G_{dt-\epsilon}/G_{dt}$ and then extended to $G/G_s \times G_{dt-\epsilon}/G_{dt}$ by defining them to be zero on $(G \cap F) \times G_{dt-\epsilon}/G_{dt}$, give all the cohomology classes in $H^1(G/G_s, \mathrm{Hom}\,(G_{dt-\epsilon}/G_{dt}, J))$.

Also, here $\lambda(l)$ stands for $\lambda + \sigma(\lambda) + \cdots + \sigma^{l-1}(\lambda)$.

*We shall be using this only for* $\epsilon = de, s = de + 1$ *and* $t = \frac{ep}{p-1} + e$.

Finally, let us note:

*Fact* 9. $H^2(G)_r = 0$ for $r < \frac{dep}{p-1}$ and $H^2(G)_{dep/(p-1)}$ constitutes the elements of order at most $p$ in $H^2(G, \mathbb{R}/\mathbb{Z})$.

Moreover, if $H^2(G, \mathbb{R}/\mathbb{Z})$ has an element of order $p^2$, then $r = \frac{dep}{p-1} + de$ where $H^2(G)_r$ is the earliest where an element of order $p^2$ shows up.

This is because the $p$-th power gives $G_{\frac{dep}{p-1}} \cong G_{\frac{dep}{p-1}+de}$ and an element $c$ of $H^2(G)_{\frac{dep}{p-1}+i}$ is of order $p^j \Leftrightarrow c^p$ in $H^2(G)_{\frac{dep}{p-1}+i-de}$ is of order $p^{j-1}$. Also, we have $p|e$ because cohomology 'pops up' at stages which are multiples of $pd$. Thus, $pd/de$ i.e. $p|e$.

## 5. Outline of proof

Here is how we obtain conditions over $f$ using the assumption that $H^2$ has $p^2$-torsion. We look at the corresponding element in

$$H^1(G/G_{1+de}, H^1(G_{dep/(p-1)}/G_{dep/(p-1)+de}, J)).$$

We consider the abelian subgroup $A = K \cap G$ of $G$ and, as elements of the above cohomology group, we may write down the equations $[X, Y^{p^2}] = 0 \ \forall\ X, Y \in \mathcal{A}$, where $\mathcal{A}$ is the image of $A$ in $G/G_{\frac{dep}{p-1}+de}$. These commutators are computed with the help of the above explicit expressions for $\epsilon = de, dt = \frac{dep}{p-1} + de, s = de + 1$ as written down in fact 8.

We note that the triviality of these commutators is due to their bilinearity. Thus we get equations over $f$ from which we try to deduce the required equations (for $p^2$th root to exist in $k$) in $f$. Note that in the computation of commutators in the central extension corresponding to $c$, the $\lambda_0$ which figures is such that $T_{r_{F/f}}(\lambda_0) \neq 0$ since $\frac{dep}{p-1} + de$ is the smallest level where $p^2$-torsion occurs.

## 6. The prime $p = 3$

We carry out the computations only in the following special case.

We assume $p = 3, d = 2, e = p(p-1) = 6$.

As $\pi^2$ is a uniformising parameter for $k$, we can write $\frac{3}{\pi^{12}} = \theta + \theta_2\pi^2 + \theta_4\pi^4 + \ldots$ where $\theta, \theta_2, \cdots \in f$.

For a $p^2$-th root of 1 to exist in $k$, it is necessary and sufficient that $\exists Y \in U_{\frac{de}{p(p-1)}}$ such that $Y^{p^2} \in U_{\frac{dep}{p-1}+de+1}$ and such that $Y \notin U_{\frac{de}{p-1}}$ i.e. $Y$ is not a $p$-th root.

In our case, we want $Y \in U_2 \setminus U_6$ such that $Y^9 \equiv 1 \bmod \pi^{31}$. Then,

$$\theta Y_2^3 + Y_2^9 = 0$$

$$\theta_2 = 0$$

$$\theta_4 = 0$$

$$\theta(Y_4 + Y_2^2)^3 + \theta_6 Y_2^3 = 0$$
$$\theta^2 Y_2 + \theta_8 Y_2^3 = 0$$
$$\theta^2 (Y_4 + Y_2^2) + \theta_{10} Y_2^3 = 0$$
$$\theta^2 Y_6 + \theta Y_6^3 + 2(\theta^2 Y_2 Y_4 + \theta Y_2^3 Y_4^3) + \theta_6(Y_4^3 + Y_2^6) + \theta_{12} Y_2^3 = 0.$$

We can find such a $Y$ if, and only if, the following 'compatibility' conditions hold

$$\theta_2 = 0$$
$$\theta_4 = 0$$
$$\theta + X^2 = 0 \text{ has a solution } \mu \text{ in } f$$
$$\theta_8^3 = \theta^5$$
$$\theta_{10}^3 + \theta^4 \theta_6 = 0$$
$$X^9 + \theta^3 X^3 = \theta_6^3 + \theta_6^6 \mu^{-9} - \theta_6 \mu^6 + \theta_{12}^3 \mu^{-3} \text{ has a solution over } f.$$

If these hold, then the solutions for $Y$ are:

$$Y_2 = \text{A solution of } Y_2^6 + \theta = 0$$
$$Y_4 = \text{A solution of } Y_4^3 = \theta_6 Y_2^{-3} - Y_2^6$$
$$Y_6 = \text{A solution of } Y_6^9 + \theta^3 Y_6^3 = \theta_6^3 + \frac{\theta_6^6}{Y_2^{27}} + \theta^3 \theta_6 + \frac{\theta_{12}}{Y_2^9}.$$

Let us expand the relevant powers of $Y$ now using the expression of $p$ in $k$.

$$Y = 1 + Y_2\pi^2 + Y_4\pi^4 + Y_6\pi^6 + \ldots\ldots \in U_2$$
$$\begin{aligned}
Y^3 &= 1 + (\theta\pi^{12} + \theta_2\pi^{14} + \theta_4\pi^{16} + \ldots)(Y_2\pi^2 + Y_4\pi^4 + Y_6\pi^6 + \ldots) \\
&\quad + (\theta\pi^{12} + \theta_2\pi^{14} + \ldots)(Y_2^2\pi^4 + 2Y_2Y_4\pi^6 + (Y_4^2 + 2Y_2Y_6)\pi^8 + \ldots) \\
&\quad + (Y_2^3\pi^6 + Y_4^3\pi^{12} + Y_6^3\pi^{18} + \ldots) \\
&= 1 + Y_2^3\pi^6 + Y_4^3\pi^{12} + \theta Y_2\pi^{14} + (\theta Y_2^2 + \theta Y_4 + \theta_2 Y_2)\pi^{16} \\
&\quad + (Y_6^3 + 2\theta Y_2 Y_4 + \theta Y_6 + \theta_2 Y_2^2 + \theta_2 Y_4 + \theta_4 Y_2)\pi^{18} + \ldots
\end{aligned}$$
$$\begin{aligned}
Y^9 &= 1 + (\theta\pi^{12} + \theta_2\pi^{14} + \ldots)(Y_2^3\pi^6 + Y_4^3\pi^{12} + \theta Y_2\pi^{14} + (\theta Y_2^2 + \theta Y_4 + \theta_2 Y_2)\pi^{16} \\
&\quad + (Y_6^3 + 2\theta Y_2 Y_4 + \theta Y_6 + \theta_2 Y_2^2 + \theta_2 Y_4 + \theta_4 Y_2)\pi^{18} + \ldots) \\
&\quad + (\theta\pi^{12} + \theta_2\pi^{14} + \ldots)(Y_2^6\pi^{12} + 2Y_2^3 Y_4^3\pi^{18} + \ldots) + (Y_2^9\pi^{18} \bmod \pi^{36}) \\
&= 1 + (\theta Y_2^3 + Y_2^9)\pi^{18} + \theta_2 Y_2^3\pi^{20} + \theta_4 Y_2^3\pi^{22} + (\theta Y_2^6 + \theta Y_4^3 + \theta_6 Y_2^3)\pi^{24} \\
&\quad + (\theta^2 Y_2 + \theta_2 Y_4^3 + \theta_2 Y_2^6 + \theta_8 Y_2^3)\pi^{26}
\end{aligned}$$

$$+ (\theta^2 Y_2^2 + \theta^2 Y_4 + 2\theta\theta_2 Y_2 + \theta_4 Y_2^6 + \theta_4 Y_4^3 + \theta_{10} Y_2^3)\pi^{28}$$
$$+ (\theta^2 Y_6 + \theta Y_6^3 + 2(\theta^2 Y_2 Y_4 + \theta Y_2^3 Y_4^3) + \theta_6 (Y_4 + Y_2^2)^3$$
$$+ \theta_{12} Y_2^3 + 2\theta\theta_2 Y_2^2 + 2\theta\theta_2 Y_4 + 2\theta\theta_4 Y_2 + \theta_2^2 Y_2)\pi^{30} + \dots .$$

Now $H^1\big(G/G_{1+de}, \mathrm{Hom}\,\big(G_{\frac{dep}{p-1}}/G_{\frac{dep}{p-1}+de}, J\big)\big) = H^1\big(G/G_{13}, \mathrm{Hom}\,(G_{18}/G_{30}, J)\big)$. Consider $X, Y \in K^1$ such that $Y^9 \in G_{18}/G_{30}$.

One knows then that $[X, Y^9] = [X, Y]^9 = 1$ and $[X, Y^9]$ can be calculated from our knowledge of $H^1(G/G_{13}, \mathrm{Hom}\,(G_{18}/G_{30}, J))$.

In fact, if $X = 1 + \sum_1^6 X_{2i}\pi^{2i}$ and $Y = 1 + \sum Y_{2i}\pi^{2i}$, if $Y^9 = 1 + \sum_9^{14} b_{2i}\pi^{2i}$, then

$$[X, Y^9] = Tr_{F/f_0}\left\{\sum_{0\le u\le 5}\ \sum_{\ell+m\le 12-2u} \lambda_u(\ell) X_\ell \bar{X}_m b_{30-2u-\ell-m}\right\}. \qquad (1)$$

Here $X^{-1} = 1 + \sum \bar{X}_{2i}\pi^{2i}$. Since $Y^9 \in G_{18}/G_{30}$, we have $b_i = 0$ for $i \neq$ 18, 20, 22, 24, 26 or 28.

Contributions to the right side of (1) are as follows:

For $u = 5$, it is

$$Tr_{F/f_0}\{\lambda_5(2) X_2 b_{18}\} = Tr_{F/f_0}\left\{Tr_{F/f}(\lambda_5) X_2(\theta Y_2^3 + Y_2^9)\right\}.$$

For $u = 4$, it is

$$Tr_{F/f_0}\left\{\lambda_4(2)(X_2 b_{20} + X_2\bar{X}_2 b_{18}) + \lambda_4(4) X_4 b_{18}\right\}$$
$$= Tr_{F/f_0}\left\{Tr_{F/f}(\lambda_4)\left[X_2\theta_2 Y_2^3 + (2X_4 + X_2\bar{X}_2)(\theta Y_2^3 + Y_2^9)\right]\right\}.$$

For $u = 3$, it is

$$Tr_{F/f_0}\left\{\lambda_3(2)(X_2 b_{22} + X_2\bar{X}_2 b_{20} + X_2\bar{X}_4 b_{18})\right.$$
$$\left. + \lambda_3(4)(X_4 b_{20} + X_4\bar{X}_2 b_{18}) + \lambda_3(6) X_6 b_{18}\right\}$$
$$= Tr_{F/f_0}\left[Tr_{F/f}(\lambda_3)\left\{X_2\theta_4 Y_2^3 + (2X_4 + X_2\bar{X}_2)\theta_2 Y_2^3\right.\right.$$
$$\left.\left. + (2X_4\bar{X}_2 + X_2\bar{X}_4)(\theta Y_2^3 + Y_2^9)\right\}\right].$$

Note that $\lambda_3(6) = 3T_{r_{F/f}}(\lambda_3) = 0$.

For $u = 2$, it is

$$Tr_{F/f_0}\left\{Tr_{F/f}(\lambda_2)\begin{bmatrix} X_2(\theta Y_2^6 + \theta Y_4^3 + \theta_6 Y_2^3) + (2X_4 + X_2\bar{X}_2)\theta_4 Y_2^3 \\ +(2X_4\bar{X}_2 + X_2\bar{X}_4)\theta_2 Y_2^3 + (X_8 + 2X_4\bar{X}_4 + X_2\bar{X}_6)(\theta Y_2^3 + Y_2^9)\end{bmatrix}\right\}.$$

For $u = 1$, it is

$$Tr_{F/f_0}\left\{ Tr_{F/f}(\lambda_1)\begin{bmatrix} X_2(\theta^2 Y_2 + \theta_2 Y_4^3 + \theta_2 Y_2^6 + \theta_8 Y_2^3) \\ +(2X_4 + X_2\bar{X}_2)(\theta Y_2^6 + \theta Y_4^3 + \theta_6 Y_2^3) \\ +(2X_4\bar{X}_2 + X_2\bar{X}_4)\theta_4 Y_2^3 + (X_8 + 2X_4\bar{X}_4 + X_2\bar{X}_6)\theta_2 Y_2^3 \\ +(2X_{10} + X_8\bar{X}_2 + 2X_4\bar{X}_6 + X_2\bar{X}_8)(\theta Y_2^3 + Y_2^9) \end{bmatrix}\right\}.$$

For $u = 0$, it is

$$Tr_{F/f_0}\left\{ Tr_{F/f}(\lambda_0)\begin{bmatrix} X_2(\theta^2 Y_2^2 + \theta^2 Y_4 + 2\theta\theta_2 Y_2 + \theta_4 Y_2^6 + \theta_4 Y_4^3 + \theta_{10} Y_2^3) \\ +(2X_4 + X_2\bar{X}_2)(\theta^2 Y_2 + \theta_2 Y_4^3 + \theta_2 Y_2^6 + \theta_8 Y_2^3) \\ +(2X_4\bar{X}_2 + X_2\bar{X}_4)(\theta Y_2^6 + \theta Y_4^3 + \theta_6 Y_2^3) \\ +(X_8 + 2X_4\bar{X}_4 + X_2\bar{X}_6)\theta_4 Y_2^3 \\ +(2X_{10} + X_8\bar{X}_2 + 2X_4\bar{X}_6 + X_2\bar{X}_8)\theta_2 Y_2^3 \\ +(2X_{10}\bar{X}_2 + X_8\bar{X}_4 + 2X_4\bar{X}_8 + X_2\bar{X}_{10})(\theta Y_2^3 + Y_2^9) \end{bmatrix}\right\}.$$

We also see (since $X$ is of norm 1 and $p = 3$) that

$$\begin{aligned} 2X_4 + X_2\bar{X}_2 &= -(X_4 + X_2^2) \\ 2X_4\bar{X}_2 + X_2\bar{X}_4 &= X_2^3 \\ 2X_{10}\bar{X}_2 + X_8\bar{X}_4 + 2X_4\bar{X}_8 + X_2\bar{X}_{10} &= -(X_4^3 + X_2^6). \end{aligned}$$

We have $0 = [X, Y^9] =$ Sum of these 6 terms corresponding to the values $u = 0, 1, 2, 3, 4, 5$.

Now, we start proving the compatibility conditions hold good.

Define $\tilde{X}$ by changing $X_{10}$ to $\tilde{X}_{10} = X_{10} + \mu$ for some $\mu$ of trace 0. Then, we can have $\tilde{X} \in K^1$ with $\tilde{X}_i = X_i$ for $i < 10$. Now,

$$\begin{aligned} 0 &= [\tilde{X}, Y^9] - [X, Y^9] \\ &= Tr_{F/f_0}\left\{Tr_{F/f}(\lambda_1)2\mu(\theta Y_2^3 + Y_2^9) + Tr_{F/f}(\lambda_0)2\mu\theta_2 Y_2^3\right\} \\ &= 4Tr_{f/f_0}\left\{Tr_{F/f}(\lambda_1)\mu(\theta Y_2^3 + Y_2^9) + Tr_{F/f}(\lambda_0)\mu\theta_2 Y_2^3\right\}. \end{aligned}$$

We have used the fact that when $d = 2$, $E \cdot E = f$.

For $Y_2 \neq 0$, since $\mu Y_2^3$ could be any arbitrary element of $f$, not depending on $Y_2^3$, we must have

$$Tr_{F/f}(\lambda_1)(\theta + Y_2^6) + Tr_{F/f}(\lambda_0)\theta_2 = 0 \;\forall\; Y_2 \in E.$$

If we take $\tilde{Y}_2$ whose square is not $Y_2^2$, then we get, on subtraction,

$$Tr_{F/f}(\lambda_1)(\tilde{Y}_2^6 - Y_2^6) = 0 \text{ i.e. } Tr_{F/f}(\lambda_1) = 0. \tag{2}$$

Therefore (as the existence of $p^2$-torsion implies that $Tr_{F/f}(\lambda_0) \neq 0$), we obtain

$$\theta_2 = 0.$$

Then, the first compatibility condition for the existence of a primitive $p^2$-th root of unity in $k$ is proved.

Let us do the same with $X_8$ now i.e. call $\tilde{X} = 1 + X_2\pi^2 + X_4\pi^4 + X_6\pi^6 + (X_8 + \mu)\pi^8 + (X_{10} + \mu X_2)\pi^{10} + \cdots \in K^1$.

$$\begin{aligned} 0 &= [\tilde{X}, Y^9] - [X, Y^9] \\ \Rightarrow 0 &= Tr_{f/f_0}\left\{Tr_{F/f}(\lambda_2)\mu(\theta Y_2^3 + Y_2^9) + Tr_{F/f}(\lambda_0)\mu\theta_4 Y_2^3\right\}. \end{aligned}$$

Changing $\mu$ to $\alpha\mu$ for any $\alpha \in f$, we get

$$0 = Tr_{F/f}(\lambda_2)(Y_2^9 + \theta Y_2^3) + Tr_{F/f}(\lambda_0)\theta_4 Y_2^3 \ \forall\ Y_2 \in E.$$

Again, as before, we will get

$$Tr_{F/f}(\lambda_2) = 0 \tag{3}$$

and therefore,

$$\theta_4 = 0. \tag{4}$$

This is the 2nd compatibility condition.

Now, putting $X_2 = 0$ in the original formula (which now consists of four terms since $Tr(\lambda_1) = 0 = Tr(\lambda_2)$), we get for all $X, Y \in E$

$$Tr_{f/f_0}\left\{Tr_{F/f}(\lambda_4)X(Y^9 + \theta Y^3) + Tr_{F/f}(\lambda_0)[X(\theta^2 Y + \theta_8 Y^3) + X^3(\theta Y^3 + Y^9)]\right\} = 0. \tag{5}$$

Again putting $X_4 = -X_2^2$ in the original formula, we get

$$Tr_{f/f_0}\left\{\begin{array}{l} Tr_{F/f}(\lambda_5)X_2(Y_2^9 + \theta Y_2^3) + Tr_{F/f}(\lambda_3)X_2^3(Y_2^9 + \theta Y_2^3) \\ + Tr_{F/f}(\lambda_0)[X_2(\theta^2 Y_2^2 + \theta^2 Y_4 + \theta_{10} Y_2^3) + X_2^3(\theta Y_2^6 + \theta Y_4^3 + \theta_6 Y_2^3)] \end{array}\right\} = 0.$$

Putting $Y_2 = 0$ will give

$$Tr_{f/f_0}\{Tr_{F/f}(\lambda_0)(\theta^2 XY + \theta X^3 Y^3)\} = 0 \ \forall\ X, Y \in E. \tag{6}$$

Since $Tr_{F/f}(\lambda_0) \neq 0$ and since $Z \mapsto \theta^2 Z + \theta Z^3$ is an endomorphism of $f$, therefore, it follows from (6) that $\exists Z \neq 0$ in $f$ such that $\theta^2 Z + \theta Z^3 = 0$. Thus, the 3rd compatibility condition has been proved.

In the above, instead of putting $Y_2 = 0$, let us, instead, put $Y_4 = -Y_2^2$ which will give $\forall\ X, y \in E$

$$Tr_{f/f_0}\left\{\begin{array}{l} Tr_{F/f}(\lambda_5)X(\theta Y^3 + Y^9) + Tr_{F/f}(\lambda_3)X^3(\theta Y^3 + Y^9) \\ + Tr_{F/f}(\lambda_0)(\theta_{10} XY^3 + \theta_6 X^3 Y^3) \end{array}\right\} = 0. \tag{7}$$

We shall use this later in the proof of the fifth compatibility condition.

Next our aim is to prove $\theta_8^3 = \theta^5$ which is the 4th condition. Now since $F$ is a quadratic extension of $f$ and $E$ is the trace zero elements in $F$, we have $E = f \cdot \alpha$ for some $\alpha \in F$ such that $\alpha^2 \in f^* \setminus (f^*)^2$. Let us expand (5) by putting $X = x\alpha$, $Y = y\alpha$ where $x, y \in f$. We get

$$0 = Tr_{f/f_0} \{\alpha^4 Tr(\lambda_4) Z y^2(\theta + y^6\alpha^6) + \alpha^2 Tr(\lambda_0)[Z(\theta^2 + \theta_8 y^2\alpha^2) \\ +\alpha^4 Z^3(\theta + y^6\alpha^6)]\} \tag{8}$$

where we have put $Z = xy$ and the traces inside are $Tr_{F/f}$.

Replacing $y$ by $y + 1$ and subtracting, we get

$$0 = Tr_{f/f_0} \begin{Bmatrix} \alpha^4 Tr(\lambda_4) Z[(1-y)\theta + \alpha^6((y+1)^8 - y^8)] \\ +\alpha^2 Tr(\lambda_0)[Z\theta_8\alpha^2(1-y) + \alpha^{10}Z^3(1-y^3))] \end{Bmatrix}.$$

Again, putting $y + 1$ for $y$ and subtracting,

$$0 = Tr_{f/f_0} \begin{Bmatrix} \alpha^4 Tr(\lambda_4) Z(-\theta - \alpha^6(1 + y^2 + y^4 + y^6))] \\ +\alpha^2 T_r(\lambda_0)(-Z\theta_8\alpha^2 - Z^3\alpha^{10}) \end{Bmatrix}.$$

Assuming that $\# f$ is large enough so that $\exists y_1, y_2$ s.t. $1 + y_1^2 + y_1^4 + y_1^6 \neq 1 + y_2^2 + y_2^4 + y_2^6$, we will have

$$0 = Tr_{f/f_0}\{\alpha^{10} Tr(\lambda_4) Z(y_1^2 + y_1^4 + y_1^6 - y_2^2 - y_2^4 - y_2^6)\} \ \forall\ Z \in f$$

and so

$$Tr_{F/f}(\lambda_4) = 0. \tag{9}$$

The equation (8) becomes
$0 = Tr_{f/f_0}\{\alpha^2 Tr(\lambda_0)[Z(\theta^2 + \theta_8 y^2\alpha^2) + \alpha^4 Z^3(\theta + y^6\alpha^6)]\}$. Putting $y + 1$ for $y$ and subtracting,

$$0 = Tr_{f/f_0}\{\alpha^2 Tr(\lambda_0)[Z\theta_8\alpha^2(1-y) + Z^3\alpha^{10}(1-y^3)]\}.$$

Putting $-y$ for $y$ and adding

$$0 = Tr_{f/f_0}\{\alpha^2 Tr(\lambda_0)[Z\theta_8\alpha^2 + Z^3\alpha^{10}]\}.$$

This makes the previous equation

$$0 = Tr_{f/f_0}\{\alpha^2 Tr(\lambda_0)[Z\theta_8\alpha^2 y + Z^3\alpha^{10}y^3]\}.$$

Replace $y$ by $y^2$ and substitute in (8) to get

$$0 = Tr_{f/f_0}\{\alpha^2 Tr(\lambda_0)(Z\theta^2 + Z^3\alpha^4\theta)\} \ \forall\ Z \in f.$$

Adding the last 2 equations

$$0 = Tr_{f/f_0}\{\alpha^2 Tr(\lambda_0)(Z\theta^2 + Z\theta_8\alpha^2 y + Z^3\alpha^4\theta + Z^3\alpha^{10}y^3)\}.$$

Write $xZ$ in place of $Z$ and then write $y = Z^2\alpha^{-2}$ to get

$$0 = Tr_{f/f_0}\{\alpha^2 Tr(\lambda_0)[x(\theta^2 Z + \theta_8 Z^3) + x^3(Z^3\alpha^4\theta + Z^9\alpha^4)]\} \ \forall\, x, Z.$$

For each $Z \in f$, $x \mapsto x(\theta^2 Z + \theta_8 Z^3) + x^3(Z^3\alpha^4\theta + Z^9\alpha^4)$ is an endomorphism of $f$. Therefore, $\forall\, Z \in f$, $\exists$ a corresponding $x \neq 0$ for which it vanishes. Take $Z$ such that $\theta Z^3 + Z^9 = 0$ which is possible by the validity of the 3rd compatibility condition. The corresponding $x$ satisfies

$$x(\theta^2 Z + \theta_8 Z^3) = 0 \text{ i.e. } \theta^2 Z + \theta_8 Z^3 = 0.$$

The 2 equations $\theta Z^3 + Z^9 = 0$ and $\theta^2 Z + \theta_8 Z^3 = 0$ give $\theta_8^3 = \theta^5$ which is the 4th compatibility condition.

Before proving the 5th condition, note that any other uniformising parameter of $K$ is $\pi^2$ times some unit $u$ of $K$ and so, a uniformising parameter of $k$ is $\pi^2$ times the norm of $u$. But, Norm(u) runs over all units of $k$ as $K$ is unramified. An easy computation shows that, by changing $\pi^2$, we may assume that $\theta_6 = \theta_{12} = 0$. The 6th compatibility condition then just reduces to the 3rd one. Also, the fifth condition becomes then $\theta_{10} = 0$.

To prove this holds, we start with (7) and proceed exactly as we did with (5). We get quite easily that $Tr_{F/f}(\lambda_5) = 0$ and $Tr_{F/f}(\lambda_3) = \frac{-\theta_6}{\theta} Tr_{F/f}(\lambda_0) = 0$. Therefore, (7) reduces to

$$0 = Tr_{f/f_0}\{Tr(\lambda_0)(\theta_{10} X Y^3)\} \ \forall\, X, Y \in E$$

which easily gives $\theta_{10} = 0$, the 5th condition.

Hence, we have proved:

**Theorem.** *Let $k$ be an extension of $\mathbb{Q}_3$ whose ramification index is 6. Also, assume that the residue field $f$ of $k$ is so large that there exist $a, b \in f$ such that $(1 + a^2)(1 + a^4) \neq (1 + b^2)(1 + b^4)$. Let $D$ be the quaternion division algebra over $k$. Suppose $H^2(SL(1, D), \mathbb{R}/\mathbb{Z})$ has an element of order 9. Then $k$ contains a primitive 9-th root of 1.*

## References


[L] M. Lazard, Groupes analytiques $p$-adiques, *Publ. Math.* IHES, Vol. **26** (1965) 5–219

[M] J. Milnor, *Introduction to algebraic K-theory*, Princeton University Press (1971)

[PR] G. Prasad and M. S. Raghunathan, Topological central extensions of $SL(1, D)$, *Invent. Math.* Vol. **92** (1988) 645–689

[R] M. S. Raghunathan, On the congruence subgroup problem, *Publ. Math.* IHES, Vol. **46** (1976) 107–161

[Ri] C. Riehm, *Amer. J. Math.* Vol. **92** (1970) 499–523

Proc. Int. Conf. – Number Theory No. 1, 2004, pp. 155–161.

# On the GCD of an Infinite Number of Integers

T. N. Venkataramana

School of Mathematics, Tata Institute of Fundamental Research, Homi Bhabha Road, Mumbai 400 005, India
e-mail : venky@math.tifr.res.in

August 2, 2002

## Introduction

In this paper, we consider the greatest common divisor (to be abbreviated gcd in the sequel) of suitable infinite sequences of integers and prove that as the sequences vary, the gcd remains bounded. These questions are motivated by results of [S] on the congruence subgroup problem. Using these results, we also obtain bounds on the indices of certain congruence subgroups of arithmetic groups in algebraic groups.

The simplest example of the result on gcd is the following. Let $a$ and $b$ be coprime positive integers and for an integer $n \neq 0$ let $\phi(n)$ be the number of positive integers not exceeding $|n|$ (the absolute value of $n$) and coprime to $|n|$. Consider the infinite sequence

$$\phi(ax+b)\ ;\ x = \cdots, -2, -1, 0, 1, 2, \cdots \tag{1}$$

and let $g(a, b)$ denote the gcd of the numbers occurring in the sequence (1). Then (see Theorem 1 below), $g(a, b)$ is bounded by 4 , for all $a$ and $b$. This assertion will be shown to be a simple application of Dirichlet's Theorem on the infinitude of primes in an arithmetic progression.

We will also prove an analogue of the above result for arbitrary number fields. Let $S$ be a finite set of places of a number field $F$ containing all the archimedean ones, $a$ and $b$ be coprime elements in the ring of $S$-integers $O_S$ , and for a nonzero ideal $\mathfrak{b}$ of $O_S$, $\phi(\mathfrak{b})$ denote the number of units in the residue class ring $O_S/\mathfrak{b}$. Write $(y)$ for the principal ideal in $O_S$ generated by the element $y$ in $O_S$. Then we show (see Theorem 2 below) that the gcd $g(a, b)$ of the infinite sequence of numbers

$$\phi((ax+b));\ x \in O_S, \tag{2}$$

is bounded by a constant depending only on the number field $F$. In fact the constant will be shown to depend only on the number of roots of unity in the number field. As in the case of $\mathbb{Q}$, Theoerm 2 will be shown to be a consequence of the Cebotarev density Theorem.

This result has an application to algebraic groups. Theorem 2 will be shown to imply that if $\Gamma$ is an $S$-arithmetic congruence subgroup of a linear algebraic group $G$ defined over the number field $F$, then the congruence subgroups $\Gamma(ax+b)$ (of $\Gamma$ for $x$ varying in $O_S$, generate a subgroup $\Gamma'$ of finite index $g$ in $\Gamma$, and the index $g$ is bounded by a constant depending only on the algebraic group $G$ and the number field $F$ (and not on the congruence subgroup $\Gamma$).

With these notations, we first prove (in §1) the following:

**Theorem 1.** *Suppose that $a$ and $b$ are coprime rational integers , and let $\phi$ denote the Euler $\phi$ function. Then the gcd $g = g(a,b)$ of the sequence $\{\phi(ax+b);\ x \in \mathbb{Z}\}$ is bounded by* 4*; more precisely, $g$ has the following values.*

(1) $g = 1$ *if* $b \equiv \pm 1 \pmod{a}$ *or* $\pm 2 \pmod{a}$.
(2) $g = 4$ *if $a$ is divisible by four, $b$ is not as in Case 1,* $b \equiv 1 \pmod{4}$ *and $b$ is not a square* $\pmod{a}$.
(3) $g = 2$ *in all other cases.*

Theorem 1 is proved by using Dirichlet's Theorem on the infinitude of primes in arithmetic progressions.

Theorem 1 can be generalised to arbitrary number fields. Let $F$ be a number field and $S$ a finite set of places of $F$ including all the archimedean ones. Let $O_S$ denote the ring of *S-integers* of $F$. Given a non-zero ideal $\mathfrak{a}$ of $O_S$, let $\phi(\mathfrak{a})$ denote the number of **units** in the quotient ring $O_S/\mathfrak{a}$ (of course, the quotient has only a finite number of elements). Given a non-zero element $x \in O_S$, denote by $\phi(x)$ the number $\phi((x))$, where $(x)$ is the principal ideal of $O_S$ generated by $x$. Recall that two elements $a$ and $b$ of $O_S$ are said to be coprime, if the ideal generated by $a$ and $b$ is the whole ring $O_S$. Let $\mu_F$ denote the group of roots of unity in the number field $F$.

**Theorem 2.** *Let $a$ and $b$ denote two elements of $O_S$ which are coprime. Let $g$ be the gcd of the numbers $\phi(ax+b)$ where $x$ runs through elements of $O_S$. Then, there is a constant $C$ depending only on the number field $K$ (and not on the elements $a$ and $b$) such that $g \leq C$. In fact, $C$ may be chosen to be Card $(\mu_F)^2$.*

*More precisely, $g$ divides Card $(\mu_F)^2$. That is $v_l(g) \leq v_l(Card(\mu_F)^2)$ where for a prime $l$ and an integer $m$, $v_l(m)$ is the largest power of $l$ which divides $m$.*

*In particular, $g$ is divisible only by the primes which divide the order of the group $\mu_F$ of roots of unity in the number field $F$.*

Theorem 2 has an application to algebraic groups. Indeed, its formulation was suggested by a problem in the arithmetic of algebraic groups. Let $G$ be a linear algebraic group defined over a number field $F$ and let $O_S$ be as before. Let $\Gamma$ be an $S$-congruence subgroup of the group $G(O_S)$ of $O_S$-rational points of $G$. Let $a$ and $b$ be coprime elements in $O_S$ and given $x \in O_S$, let $G(ax+b)$ be the congruence subgroup corresponding to the principal ideal $(ax+b)$ of $O_S$. Let $\Gamma'$ be the subgroup of $\Gamma$ generated by the subgroups $\Gamma \cap G(ax+b)$ as $x$ varies through elements of the ring $O_S$.

**Theorem 3.** *There exists an integer $g$ depending only on the group $G$ and the number field $F$ such that for all $\Gamma$ as above, the index of $\Gamma'$ in $\Gamma$ is bounded by $g$.*

## 1. Proof of theorem 1

As we said before, the proof is a repeated application of Dirichlet's Theorem on the infinitude of primes in arithmetic progressions. We will prove it by a (small) number of case by case checks.

[1] Suppose that $b$ is congruent to either $\pm 1$ or $\pm 2$ modulo $a$. Then, there exists an integer $x$ such that $ax + b$ is either $\pm 1$ or $\pm 2$. In either case $\phi(ax + b) = 1$; since the gcd $g$ divides each $\phi(ax + b)$, we obtain that $g = 1$.

Let $l$ be a prime and $m$ an integer. Define, as before, $v_l(m)$ to be the largest integer $k$ such that $l^k$ divides $m$.

[2] We first prove that

$$v_l(g) = 0$$

if $l$ is an odd prime. Write $a'$ for the l.c.m. of $a$ and $l$.

If $b$ is coprime to $l$ and $b \not\equiv 1 \pmod{l}$, then, by Dirichlet's Theorem, the residue class $b$ modulo $a'$ is represented by a prime $p$. Thus, there exist integers $x, y$ such that $p = a'y + b = ax + b$, and $\phi(ax + b) = \phi(p) = p - 1 \not\equiv 0 \pmod{l}$. Therefore, $g$ (which divides $\phi(p)$) is coprime to $l$ as well. That is $v_l(g) = 0$.

If $b$ is not coprime to $l$, then, by replacing $b$ by $a + b$ we may still assume that $b$ is coprime to $l$, as $a$ and $b$ are coprime. Suppose now that $b \equiv 1 \pmod{l}$. Since the group $(\mathbb{Z}/l\mathbb{Z})^*$ of units in the ring $\mathbb{Z}/l\mathbb{Z}$ has at least two elements, and $(\mathbb{Z}/a'\mathbb{Z})^* \to (\mathbb{Z}/l\mathbb{Z})^*$ is surjective, we may write $b = uv$ modulo $a'$ with $u \not\equiv 1$ and $v \not\equiv 1$ modulo the prime $l$. Moreover, by Dirichlet's Theorem, $u$ and $v$ modulo $a'$ are represented by (distinct) primes $p$ and $q$. Thus, there exist $x$ and $y$ with $uv = a'y + b = ax + b$ and $p \equiv u$ and $q \equiv v$ modulo $a'$. Moreover, $p - 1 \not\equiv 0 \pmod{l}$ and $q - 1 \not\equiv 0 \pmod{l}$. Thus, $\phi(pq) = (p - 1)(q - 1) \not\equiv 0 \pmod{l}$ whence $v_l(g) = 0$. This completes [2].

[3] Now we consider the case when $l$ is even ($l = 2$). We show that

$$v_2(g) \leq 2.$$

**(3. 1)** Suppose that $a$ is even and that $a^*$ is the lcm of $a$ and 4. Now, $b$ is odd since $b$ is coprime to $a$. Thus $b$ defines a class in $(\mathbb{Z}/4\mathbb{Z})^*$.

If $b$ is a non-trivial class in $(\mathbb{Z}/4\mathbb{Z})^*$, then, since the class of $b$ modulo $a^*$ is represented by a prime $p$, one has $p = ax + b$ for some $x$ with $\phi(ax + b) = \phi(p) = p - 1 \not\equiv 0 \pmod{4}$. Thus, $v_2(\phi(ax + b)) \leq 1$ whence $v_2(g) \leq 1$.

If $b$ *is* the trivial class modulo 4, then write $b \equiv (-b)(-1) \pmod{a^*}$. Both $(-b)$ and $(-1)$ are represented by primes $p$ and $q$ say. Then $p \equiv q \equiv (-1) \pmod{4}$, whence, $p - 1$ and $q - 1$ are divisible at most by 2. Thus, there exists an $x$ with $ax + b = pq$, with $\phi(ax + b) = \phi(p)\phi(q) = (p - 1)(q - 1) \not\equiv 0 \pmod{8}$. Hence $v_2(g) \leq v_2(\phi(ax + b)) \leq 2$.

**(3. 2)** Suppose $a$ is odd. Let $a^* = 4a$ denote the lcm of $a$ and 4. The group $(\mathbb{Z}/a^*\mathbb{Z})^*$ of units is a product of $(\mathbb{Z}/a\mathbb{Z})^*$ and $(\mathbb{Z}/4\mathbb{Z})^*$. Hence there exists a prime $p$ such that it represents $b$ modulo $a$ and $-1$ modulo 4. Thus there exists an $x$ with $p = ax + b$. Moreover $\phi(ax + b) = \phi(p) = p - 1 \not\equiv 0 \pmod{4}$, whence $v_2(g) \leq 1$.

[4] Suppose that $ax + b$ is at least 3 in absolute value. Assume also that $a \equiv 0 \pmod 4$, $b = 1 \pmod 4$ and that $b$ is not a square (mod $a$). Write $ax + b = p_1^{e_1} \cdots p_h^{e_h}$, with $p_i$ distinct primes and $e_i$ positive integers. Clearly, the primes $p_i$ are all odd.

If $h \geq 2$, then since $\phi(ax+b)$ is divisible by the product $(p_i - 1) \cdots (p_h - 1)$, it follows that $\phi(ax + b)$ is divisible by $2^h$ and hence by 4.

If $h = 1$ and $p_1 = 3 \pmod 4$, then by passing to congruences modulo 4 (note that 4 divides $a$ by assumption), we see that $1 \equiv b \equiv p_1^{e_1} \equiv (-1)^{e_1}$ modulo 4, whence $e_1$ is even, contradicting the assumption that $b$ is not a square modulo $a$.

If $h = 1$ and $p_1 \equiv 1 \pmod 4$, then $\phi(ax + b) = p_1^{e_1 - 1}(p_1 - 1)$ is divisible modulo 4.

We have thus proved part (2) of Theorem 1, that for all $x$, $\phi(ax+b)$ is divisible by 4; that is, 4 divides $g$. Since we have proved in all cases that $v_2(g) \leq 2$, it follows that $g = 4$ if the conditions of Part (2) hold.

[5] Turn now to the other parts of Theorem 1. If $a \equiv 2 \pmod 4$, then write $a = 2a_1$, with $a_1$ odd. By Dirichlet's Theorem, there exist infinitely many primes $p$ with $p \equiv b \pmod{a_1}$ and $p \equiv 3 \pmod 4$. Note that $b \equiv p \equiv 1 \pmod 2$ since both $b$ and $p$ are odd. Thus, $p \equiv b \pmod a$ and $\phi(p) = p - 1 \not\equiv 0 \pmod 4$ whence $v_2(g) \leq 1$. Since $g$ is always even in this case, it follows that $g = 2$.

If $b$ is $\pm 1$ or $\pm 2$, then, there exists an integer $x$ such that $ax + b = \epsilon$ with $\epsilon = \pm 1$ or $\epsilon = \pm 2$. Clearly, then $\phi(\epsilon) = 1$ whence $g = 1$.

If $a$ and $b$ are as in Part (3) of Theorem 1, then $ax + b$ is at least 3 in absolute value for all $x$ and hence $\phi(ax+b) \geq 2$. If $a$ is divisible by 4, but $b$ is a square say $c^2$, modulo $a$, then $c$ is represented by a prime $p$ modulo 4 which (by replacing $c$ by $-c$ if necessary) may be assumed to be congruent to $-1$ modulo 4. Hence $p^2 = ax + b$ for some integer $x$, and $\phi(ax + b) = p(p - 1) \not\equiv 0 \pmod 4$. Therefore, $v_2(g) \leq 1$ whence $g = 2$. The rest of Theorem 1 may be proved in exactly the same way.

## 2. Proof of theorem 2

### 2.1 *Preliminaries*

Denote by $\mu$ the order of the group of roots of unity in the number field $F$. Let $l$ be a prime and $l^e$ the largest power of $l$ which divides $\mu$. Let $\omega$ be a primitive $l^{e+1}$-th root of unity in an algebraic closure of $F$. The Galois group of the extension $F(\omega)/F$ is a subgroup of the group $(\mathbb{Z}/l^{e+1}\mathbb{Z})^*$ of units and is non-trivial, since $l^e$ is the highest power of $l$ which divides $\mu$. Moreover, if $\mathfrak{p}$ is a prime ideal in $O_S$ which is unramified in $F(\omega)$ such that its Frobenius is a non-trivial element in $Gal(F(\omega)/F)$ then, $N(\mathfrak{p}) - 1$ is divisible by $l^e$ but not by $l^{e+1}$.

### 2.2 *Notation*

Denote by $F_a$ the ray class field of $F$ corresponding to the ideal $O_S$. Thus, by definition, $F_a/F$ is an abelian extension whose Galois group is isomorphic -under the Artin

Reciprocity map- to the quotient $G(F_a/F) = \mathbb{A}_F^*/F^* \left(\prod U_v\right)$. Here for a commutative ring $R$, we denote by $R^*$ the group of units of $R$ and $A_F$ is the ring of adeles of $F$. For each finite place $v$ of $F$, let $O_v$ be the maximal compact subring of the completion $F_v$. If $v \in S$, let $U_v = K_v^*$ and if $v \notin S$, let $U_v$ denote the subgroup of elements of $O_v^*$ congruent to 1 modulo the ideal of $O_v$ generated by $a$. Note that $S$ contains all the archimedean primes and that if $v \notin S$ and if $v$ does not divide $a$, then $U_v = O_v^*$.

Given an element $z = \prod(z_v)_v$ of $A_F^*$, define its **norm** by $\prod(|z|_v)$ for all $v \notin S$. In particular, its restriction to $\prod O_v^*$ (where $v$ runs through all finite places of $F$) is trivial, and hence the norm is a homomorphism on the group of fractional ideals of $O_S^*$. Thus, if $\mathfrak{p}$ is a prime ideal in $O_S$ and $v$ the corresponding non-archimedean absolute value on $F$, then the norm is 1/Card $(O_S/\mathfrak{p})$.

By the weak approximation theorem, it is clear that the group

$$G(F_a/F) = \mathbb{A}_F^*/F^* \left(\prod U_v\right)$$

is isomorphic to the group

$$\mathbb{A}_F(S)^*/O_S(a)^* \left(\prod O_v^*\right)$$

where $O_S(a)^*$ is the group of units in the ring $O_S$ which are congruent to the identity modulo the principal ideal $a$ of $O_S$ and the product is over all the places $v$ of $F$ not in $S$ and not dividing $a$.

### 2.3 *Proof of theorem 2*

Let $E$ be the compositum of the ray class field $F_a$ and the cyclotomic extension $F(\omega)/F$ ($\omega$ being a primitive $L^{e+1}$-th root of unity). The Galois group $G(E/F)$ is abelian and has surjective homomorphisms onto $G(F_a/F)$ and onto $G(F(\omega)/F)$. The group $G(E/F)$ is generated-by the Cebotarev Density Theorem-by the Frobenius elements corresponding to prime ideals $\mathfrak{p}$ of the ring $O_S$ (which may also be assumed to be unramified in the extension $E/F$). In particular, the image of $b \in F$ ($F^*$ thought of as a subgroup of $\mathbb{A}_F(S)^*$ (= the group of $S$-ideles)) in $G(E/F)$ is represented by (infinitely many) prime ideals $\mathfrak{p}$. This means that $\mathfrak{p} = (ax + b)O_S$ for some $x \in O_S$. In particular, $\phi(\mathfrak{p}) = \phi(ax + b)$.

*Case 1.* If moreover, there exists an element $\sigma \in G(E/F)$ such that its image in $G(F_a/F)$ is represented by the element $b$ and its image in $G(F(\omega)/F)$ is non-trivial, then $\sigma$ is represented by a prime ideal $\mathfrak{p}$, with the following properties. By the choice of $\sigma$, one gets an $x \in O_S$ such that $ax + b = \mathfrak{p}$. Thus $\phi(ax + b) = \phi(\mathfrak{p})$ and $\mathfrak{p}$ has non-trivial image in $G(F(\omega)/F)$. Hence $\phi(\mathfrak{p})$ is divisible by $l^e$ but not by $l^{e+1}$ (see (2.1)).

*Case 2.* If Case 1 is not possible, then this means that the preimage of $b \in G(F_a/F)$ in $G(E/F)$ maps to the trivial element in $G(F(\omega)/F)$. Fix any $\sigma$ in the preimage, and let $\tau \in G(E/F)$ map non-trivially into $G(F(\omega)/F)$ (recall from (2.1) that the latter group is non-trivial). One may write $\sigma = (\sigma\tau)(\tau)^{-1}$. Note that both $\sigma\tau$ and $\tau^{-1}$ have non-trivial images in $G(F(\omega)/F)$. Represent both elements by prime ideals $\mathfrak{p}$ and $\mathfrak{q}$

respectively. Then $\phi(\mathfrak{p})$ is divisible by $l^e$ but not by $l^{e+1}$ (similarly for $\phi(\mathfrak{q})$). However, $\mathfrak{p}\mathfrak{q} = ax + b$ for some $x \in O_S$. Therefore, $\phi(ax + b)$ is divisible by $l^{2e}$ but by no higher power of $l$. In particular, $v_l(g) \leq 2e$.

The combination of Cases 1 and 2 proves Theorem 2.

## 3. A question

In sections 1 and 2 we considered the gcd of $\phi(ax + b)$ as $x$ varies. Here $ax + b$ is a **linear** polynomial of degree one, with coprime coefficients. We now replace $ax + b$ by a polynomial $P(x) = a_n x^n + \cdots + a_0$ where $a_0, \cdots, a_n$ are integers (or S-integers) which are "coprime" i.e. generate the unit ideal. We have the following question, arrived at in conversations with M. S. Raghunathan.

3.1 *Question*

Does There exist a constant $C = C(n)$ depending only on the degree $n$ of the polynomial $P$ such that the gcd of the integers $\phi(P(x)) : x = 0, 1, \cdots$ is bounded by $C(n)$ for **all** polynomials $P$ of degree $n$ with coprime coefficients ?

In case the number field $F$ is different from $\mathbb{Q}$, the question amounts to asking if the constant $C$ depends only on the number field $F$ and on the finite set $S$ of places.

## 4. Relation to the congruence subgroup problem

Consider the group $SL_2$ over a number field $F$. Suppose the set $S$ is such that the group $O_S^*$ of units is infinite. Given an ideal $\mathfrak{a}$ of $O_S$, let $G(\mathfrak{a})$ denote the principal congruence subgroup of $SL_2(O_S)$ of level $\mathfrak{a}$: this is the set of matrices of determinant one, with entries in the ring $O_S$ which are congruent to the identity matrix modulo the ideal $\mathfrak{a}$. Let $E(\mathfrak{a})$ denote the normal subgroup of $SL_2(O_S)$ generated by the upper triangular matrices with 1's on the diagonal and which are congruent to 1 modulo the ideal $\mathfrak{a}$. Then, $E(\mathfrak{a})$ is a normal subgroup of $G(\mathfrak{a})$.

In the course of proof of the congruence subgroup property for the group $SL_2(O_S)$, Serre (in [S]) considers the action of the group $T(O_S) \simeq O_S^*$ on the quotient $G(\mathfrak{a})/E(\mathfrak{a})$. Here $T$ is the group of diagonals in $SL_2$. Serre shows that this action is trivial on the subgroup of $T(O_S)$ generated by $\mu$-th powers of elements of $T(O_S)$ (recall that $\mu$ is the number of roots of unity in the number field $F$). This is achieved by proving first that if a coset class $\xi$ in the quotient $SL_2(\mathfrak{a})/E(\mathfrak{a})$ is represented by a matrix $A = \left(\begin{smallmatrix} a & b \\ c & d \end{smallmatrix}\right)$, then, the group $T(aO_S)$ acts trivially on the element $\xi$. Next, if the element $A$ is replaced by $BA$ with $B = \left(\begin{smallmatrix} 1 & x \\ 0 & 1 \end{smallmatrix}\right)$ with $x \in \mathfrak{a}$, then $BA = \left(\begin{smallmatrix} a' & b' \\ c' & d' \end{smallmatrix}\right)$ with $a' = ax + b$. Consequently, even the group $T((ax + b)O_S)$ acts trivially on the element $\xi$. Thus the group $T_0$ generated by the subgroups $T((ax + b)O_S)$, as $x$ varies, acts trivially on the element $\xi$. By arguments similar to the proof of Theorem 2 (indeed, the proof of Theorem 2 in the present paper is modelled on [S]), Serre then shows that the group of $\mu$-th powers of $T(O_S)$ is contained in the group $T_0$ and hence acts trivially on $\xi$.

We now state a generalisation of this result. Let $G \subset GL_n$ be a linear algebraic group defined over a number field $F$. Let $S$ be a finite set of places of $F$ containing all the archimedean ones and let $O_S$ denote the ring of $S$-integers of $F$. Given a non-zero ideal $\mathfrak{a}$ in $O_S$, denote by $G(\mathfrak{a})$ the subgroup of $G \cap GL_n(O_S)$ whose entries are congruent to the identity modulo the ideal $\mathfrak{a}$.

Suppose now that $a$ and $b$ are coprime elements in $O_S$ and that $x \in O_S$. Denote, as before, by $G(ax+b)$, the principal congruence subgroup of level $ax+b$. Let $\Gamma \subset G(F)$ be an arithmetic subgroup and let $\Gamma_{a,b}$ be the group generated by the subgroups $G(ax+b) \cap \Gamma$ for all $x \in O_S$ of $G(O_S)$.

**Theorem 4.** *With the foregoing notation, the group $\Gamma_{a,b}$ has its index in $\Gamma$ bounded by a constant dependent only on $G \subset GL_n$ and the number field $F$, but neither on the arithmetic group $\Gamma$ nor on the elements $a$ and $b$.*

We begin by noting that for the multiplicative group $G = \mathbf{G}_m$, this is exactly Serre's Theorem: the index is bounded by $\mu$, the number of roots of unity in the number field.

The proof of Theorem 4 is easy when $G$ is unipotent or semi-simple, since one may appeal to the strong approximation theorem, a version of which implies that if $\mathfrak{a}$ and $\mathfrak{b}$ are coprime ideals, then $G(\mathfrak{a})$ and $G(\mathfrak{b})$ generate a subgroup of index bounded by a constant independent of $\mathfrak{a}$ and $\mathfrak{b}$ (see [R], section 2 for similar considerations). Thus, the proof of Theorem 3 may be reduced easily to that for tori, and hence, by the structure theorem for tori , to that for the groups $\mathbf{G}_m$ over some *finite extensions* $E$ of $F$. Thus, we are reduced to the case of $\mathbf{G}_m$ over number fields, for which Serre's result is already available.

## 5. Acknowledgements

The author is grateful to S. D. Adhikari, R. Balasubramanian and to K. Srinivas for their kind invitation to take part and to contribute to the proceedings of a conference in honour of Prof.Subba Rao held at Institute of Mathematical Sciences, Chennai. He is also grateful to the referee for a careful reading of the MS, for making several necessary corrections, and for pointing out a gap (which is now filled!) in the proof of Theorem 1.

The author acknowledges the help of M. S. Raghunathan in arriving at the question raised in section 3. He is also grateful to J.-P. Serre for suggesting a possible solution through sieve theoretic methods.

## References

[S] J.-P. Serre, Le Probleme de groupes de congruence pour $SL_2$, *Ann. Math.* **92** (1970) 489–527

[R] M. S. Raghunathan, On the Congruence Subgroup Problem II, *Invent.Math.* **85** (1986) 73–117

[V] T. N. Venkataramana, On Systems of Generators of Higher Rank Arithmetic Subgroups, *Pac. J. Math.* **166** no. 1 (1994) 193–212

Made in United States
North Haven, CT
05 November 2026

21778845R10096